OPTICAL PROPERTIES OF THIN SOLID FILMS

OPTICAL PROPERTIES
OF THIN SOLID FILMS

O. S. HEAVENS, Ph.D.

*Lecturer in Physics at
the University of Reading*

LONDON
BUTTERWORTHS SCIENTIFIC PUBLICATIONS
1955

BUTTERWORTHS PUBLICATIONS LTD.
88 KINGSWAY, LONDON, W.C.2

AFRICA: BUTTERWORTH & CO. (AFRICA) LTD.
                     DURBAN: 33/35 Beach Grove

AUSTRALIA: BUTTERWORTH & CO. (AUSTRALIA) LTD.
                     SYDNEY: 8 O'Connell Street
                     MELBOURNE: 430 Bourke Street
                     BRISBANE: 240 Queen Street

CANADA: BUTTERWORTH & CO. (CANADA) LTD.
                     TORONTO: 1367 Danforth Avenue

NEW ZEALAND: BUTTERWORTH & CO. (AUSTRALIA) LTD.
                     WELLINGTON: 49/51 Ballance Street
                     AUCKLAND: 35 High Street

U.S.A. Edition published by
ACADEMIC PRESS INC., PUBLISHERS
125 EAST 23RD STREET
NEW YORK 10, NEW YORK

Set in Monotype Baskerville type
Made and printed in Great Britain by William Clowes and Sons, Limited
London and Beccles

PREFACE

THE earliest recorded scientific studies on the optical behaviour of thin films were made in the seventeenth century. They appear, together with the results of a wide range of other optical experiments, in Sir Isaac Newton's early treatise on Opticks, written in a style which has a charm and fascination of its own. It is a tribute to Newton that reference to his work needs to be made (and not merely from the standpoint of historical interest) in a book written in the twentieth century. Although the mechanism of the propagation of light was in Newton's time shrouded in uncertainty, the observations made on the colours of thin films in relation to their thickness have proved useful for a long time and have formed the basis of certain of the methods of thickness measurement used at the present time.

With the development in recent years of methods of preparing thin films of materials, interest in the optical properties of films has been considerably stimulated. The wealth of information which has now become available is such as to deserve inclusion in a single volume. Although there remain many unsolved problems, the general features of the optical behaviour of thin films are now reasonably well understood. Progress in the methods of studying the structures of films continues apace, further assisting in the problems of interpretation of optical phenomena exhibited by films while developments in the field of high-speed computing are doing much to reduce the labour involved in studying the properties of multiple-film systems. The closing chapter of this book is devoted to a study of the practical applications of thin films, a field in which considerable and striking progress has been made.

My grateful thanks are due to Professor R. W. Ditchburn for his help and encouragement and to Mr. R. A. Hyman for his assistance with proofs. I am also indebted to Dr. F. Abelès (Paris), to Professor P. Rouard and Dr. D. Malé (Marseille) and to Dr. A. F. Turner (New York) for their help, in spite of the large distances which separate us; and to my wife for her patient help in the preparation of the copy.

Acknowledgments are also due to the proprietors of the following journals for permission to reproduce the figures indicated:—

Journal of Applied Physics, vol. 14 (Table 3.1).
Optik, vol. 7 (Fig. 3.9).

Journal of Scientific Instruments, vol. 29 (Fig. 5.1).
Canadian Journal of Research, vol. 26 (Fig. 5.12).
Journal of Chemical Physics, vol. 17 (Table 5.1).
Philosophical Magazine, vol. XLI (Fig. 5.24).
Journal of the Optical Society of America, vol. 40 (Fig. 6.9).

and to the Bausch & Lomb Optical Co. for permission to reproduce Fig. 7.8.

O. S. HEAVENS

December 1954

CONTENTS

		Page
PREFACE		v
1. INTRODUCTION		1
2. THE FORMATION OF THIN FILMS		6
3. THE STRUCTURE OF SOLID FILMS		24
4. THIN FILM OPTICS		46
5. MEASUREMENTS OF FILM THICKNESS AND OPTICAL CONSTANTS		96
6. RESULTS OF OPTICAL MEASUREMENTS ON FILMS . .		155
7. PRACTICAL APPLICATIONS OF THIN FILMS IN OPTICS . .		207
INDEX		257

vii

INTRODUCTION

Growth of the importance of thin films—Methods of film deposition—Calculations of the optical properties of thin film systems—Measurements on thin films—The uses of thin films

1.1. GROWTH OF THE IMPORTANCE OF THIN FILMS

FOR nearly the first half of this century, interest in the optical properties of thin films was largely confined to the use of reflecting films in interferometry. Their importance in this connection was considerable. The high resolving powers attainable in such instruments as the Fabry-Pérot interferometer enabled an impressive accuracy to be reached in spectroscopic work and in consequence great progress was made in this field. The role played by the films was, however, a purely utilitarian one. Little attention was paid to the study of thin films for their intrinsic interest.

The results of early studies of the optical behaviour of films showed large differences from those obtained on bulk materials. This was especially true of absorbing materials. The differences were much greater than could be accounted for by considering the limitation of the extent of the material in one direction. Thin film behaviour was at that time acknowledged to be anomalous but the means necessary to enable an explanation of the anomalies to be found were not then available. Within the last two decades, interest in the optical and other properties of thin films has grown considerably. Developments of the techniques of producing and studying thin films have enabled a fairly clear picture of the nature of such films to be obtained and have led to an understanding of their optical behaviour.

The increasing use and study of thin films are in part due to the rapid strides which have been made in vacuum technique and to the development of electron-optical methods of examination. The attainment of the low pressures necessary for the production and study of films has been so facilitated that rapid progress has been made possible. Electron microscopy has enabled information to be obtained directly which could otherwise be inferred only with great difficulty and uncertainty. From the electron-optical evidence, the thin film is seen to be in a somewhat disordered state, except in certain special circumstances. The low state of order accounts for many of the apparent anomalies observed in the behaviour of thin films and poses a severe problem in the development of a theoretical approach.

1.2. METHODS OF FILM DEPOSITION

By far the most widely used method of depositing films, particularly for use in optical systems, is that of thermal evaporation and it is to this method that most attention is given in the next chapter. With no other method can so complete a measure of control be obtained. By the use of suitable shutters and diaphragms and by subtle movement of the target during the deposition of the film, practically any distribution of material on a surface may be obtained. The immediate application which springs to mind in this connection is that of producing the various aspheric surfaces which we know from geometrical optical studies to be so useful. Many such forms are extremely difficult to produce by methods other than the evaporation process—so difficult as to be of no practical interest. These and other applications of this type are discussed in the concluding chapter.

The multiplicity of variables attending the deposition process, by which such effective control may be obtained, is in some measure responsible for the lack of concordance which is so often evident in the results of thin film investigations made by different observers. Until an appreciation of the dependence of film properties on the conditions of formation of the film was obtained, little attention was directed to keeping the many important variables under control. Among the factors which are liable to influence the properties of a film are: rate of deposition, velocity of impinging atoms, structure and condition of target surface, history of film between deposition and examination, etc. Lack of information on these and the many other variables which may determine the nature of the film formed makes comparison of different workers' results very difficult. Electron-optical methods have enabled an appraisal to be made of the effect of the evaporation conditions on the structure and properties of films produced. We may now hope, therefore, for a greater degree of consistency among thin film results from different sources.

1.3. CALCULATIONS OF THE OPTICAL PROPERTIES OF THIN FILM SYSTEMS

When a beam of light traverses a stratified medium in which there are discontinuous changes in refractive index (or changes which take place in a distance small compared with the light wavelength) multiple reflections occur. If the distances between boundaries are small so that the multiply-reflected beams are coherent with one another, then the intensity of light reflected or transmitted by the system is obtained from the algebraic sum of the amplitudes. The amplitudes to be summed are calculated from Maxwell's equations with the application of the appropriate boundary conditions. For a single film bounded by surfaces of known reflectance, the reflectance and transmittance are given by the Airy summation; this is the familiar case of the Fabry-Pérot etalon. The treatment is approximate inasmuch as it is assumed that the change of phase on reflection of the light at either side of the reflecting surfaces of the etalon is the same. In practice, where a metal reflecting surface is used, this is not true. The error is small, however, for the case where high reflectances are used and this is the usual arrangement in an interferometer. If the properties of the etalon are to be calculated from the optical constants of all the layers, instead of considering the silver layers as possessing an effective reflectance, then the problem becomes very complicated. This, however, is a case of the general problem of calculating the optical properties of a set of films, from given values of the optical constants and thicknesses.

A moment's consideration shows that the method of counting up and summing the multiple reflections is likely to be of limited appeal for systems of many layers. More elegant ways are possible. The summation may be assumed to have been effected before we arrive on the scene so that we deal with the resultant amplitude of the wave in each medium. Expressions for the reflectance and transmittance are then readily found by the application of boundary conditions to the resultant waves. In principle, the solution for any number of layers is straightforward inasmuch as the reflectance and transmittance may be calculated explicitly in terms of the parameters of the system. In practice, the expressions for R and T for more than two layers are very cumbersome or intolerably cumbersome, depending on whether the layers are transparent or absorbing. The effect of inserting complex values of the Fresnel coefficients in the innocent-looking expressions for R and T for even a single film is striking, in an unpleasant way.

The methods used in dealing with this problem are dealt with in Chapter 4. The treatments given are those which can be applied without extensive higher mathematical equipment. Although in certain instances, more elegant treatments and more compact forms are possible, these generally require a considerably greater degree of mathematical skill than that possessed by the majority of physicists.

1.4. Measurements on Thin Films

With the increasing importance of thin film work, the need has arisen for information on the properties of films and this has necessitated the development of methods of investigation suitable for dealing with thin layers. Many early methods of measuring film thickness required a knowledge of the refractive index of the film and vice versa. Procedures now developed which enable both quantities to be determined independently have shown how great is the need for methods which are independent of the assumption that any of the properties of a film (save, perhaps, its chemical composition) are the same as those of the material in bulk.

The greatest difficulties arise in connection with films of absorbing materials and especially with those in which the absorption is high (metals). Unequivocal determination of the optical constants of such films is possible only by making measurements of the amplitude and phase of light beams transmitted by and reflected from both sides of the film. These measurements are difficult to make with high accuracy. The pitfalls of earlier methods, which were unable to give the optical constants and thickness simultaneously, are well illustrated by the alarming ranges covered by the various reported values of the optical constants. Such differences may in part have arisen from the use of unsuitable methods of measurement and in part from the fact that the optical constants of films may vary considerably with the conditions of preparation.

The violent variations with thickness of the optical constants of thin metal films have been shown to be a consequence of the aggregated nature of the films. The structure of metal films deduced from observations of the optical properties shows remarkable agreement with that observed directly by electron microscopy or deduced indirectly from electron diffraction and other experiments.

1.5. The Uses of Thin Films

With the rapid development of the design of demountable vacuum systems, the number and diversity of uses of thin films have increased apace. The small vacuum evaporation plant has become an almost essential part of the furniture of the research laboratory. On an industrial scale, the vacuum evaporation system has joined the ranks of routine industrial processes even to the point of being incorporated in a continuous belt system for lens blooming. (Paradoxically, it is now more difficult for a lens-producing plant to produce an unbloomed lens than a bloomed one, the greater hardness of the

4

bloomed surface making the bloomed product less susceptible to scratches than is the raw glass.)

In the laboratory, thin films find application in a wide variety of types of work. Electrical experiments may often be facilitated by the use of evaporated electrodes, which make intimate contact with the surface without causing mechanical damage. The hygroscopic optical components used in infra-red spectroscopy may be protected by coating with a suitable insoluble film. Reflecting surfaces may be similarly protected against deterioration by the atmosphere. Electrostatic charging, so often the bugbear of experiments employing fine suspensions, may sometimes be reduced by metallizing the offending components; the same process applied to the fluorescent screen of a cathode-ray tube both eliminates charging troubles and results in a brighter picture. The difficulties of measuring surface temperatures are largely overcome by the use of evaporated films as thermocouples whilst the bulk of optical measurements in the near infra-red depend on the photoconducting properties of layers produced by thermal evaporation (with a certain amount of additional persuasion).

Perhaps the most striking of the many developments have been in the field of multilayer filters. From the simple low-reflecting low-index layer and high-reflecting high-index layer, magnificently complicated multilayer systems have been evolved possessing impressive and useful optical properties. Narrow transmission bands, wide bands with steep edges, low- and high-pass filters may now be almost tailor-made to suit requirements. The fact that for some filters large numbers of layers are called for is of no consequence since the development of techniques of deposition and of monitoring have not lagged behind the theoretical work.

From the large number of papers which have appeared in the last few years on the optical properties of thin solid films, it is clear that an exhaustive treatise must be of encyclopaedic proportions. In this work examples have been drawn from a field in which progress is still rapid; they give a representative picture both of the potentialities and of the limitations of thin films in relation to optical studies.

THE FORMATION OF THIN FILMS

Introduction—The thermal evaporation process—Practical methods of evaporation—The method of sputtering—Other methods of film deposition—Fundamental aspects of the condensation process

2.1. INTRODUCTION

WE shall consider in some detail the experimental aspects of the various methods of preparing films.   Most attention is paid to the method of thermal evaporation since this method has been most extensively developed.   It possesses many advantages over the other methods described below, not the least of which lies in the ease with which the process may be controlled.   Films of high purity are readily produced and with a minimum of interfering conditions.   In spite of this favourable aspect, however, results of studies of the properties of evaporated films have not always shown the consistency which would be expected.   The cause of these discrepancies frequently lies not in any inherent difficulties of the method itself but more probably in a lack of control of the many experimental variables on the part of different investigators, who frequently fail to state the precise conditions under which films are prepared.

The evaporation method is not universally applicable in practice since, for high melting-point materials, it may well be impossible to deposit films without undue heating of the receiving substrate.   The temperature of the substrate during deposition plays an important role in determining film structure so that any method in which this factor cannot be accurately controlled is to be avoided if reproducible results are to be obtained.   For metals for which the evaporation method is unsuitable, the method of sputtering is often used.   In this method, the mean source temperature at which an appreciable rate of sputtering occurs is considerably below that at which thermal evaporation at the same rate would occur so that the amount of radiation falling on the target surface is considerably less.   The method of sputtering was used extensively before the development of thermal evaporation techniques, much of the early work on the fundamentals of film deposition (Section 2.6) having been carried out with sputtered materials.   The current tendency is to use the evaporation method where this can be applied rather than that of sputtering on account of the greater control of experimental conditions which is available. Also the possible influence of the considerable quantity of gas which requires to be present in the sputtering process is avoided by using the evaporation method. The method developed recently of sputtering in a magnetic field enables sputtering to be effected at much lower pressures than before.

The importance of the remaining methods quoted is mainly of a utilitarian nature. Little work has been done in studying the properties of electrolytically or chemically deposited films except in so far as the films are of some practical significance. For example, the optical behaviour of certain chemically deposited films has been examined since it is found to be possible to make non-reflecting films and highly-reflecting films by this method (Sections 7.2–3).

2.2. The Thermal Evaporation Process

We first consider the conditions which must be realized in order that the deposition of a film from the vapour may take place with a minimum of interfering factors. The variables which are to be expected to have some influence on the nature and properties of an evaporated film may be summarized thus:

(i) Nature and pressure of residual gas in chamber

(ii) Intensity of beam of atoms condensing on surface

(iii) Nature and condition of target surface

(iv) Temperature of evaporation source and hence velocity of impinging atoms

(v) Contamination by evaporated material by supporting material of source.

The residual gas

It is sometimes stated that the condition to be met for the deposition of films by the evaporation method is that the mean free path of the volatilizing atom shall be large compared with the dimensions of the system or, more specifically, with the source to target distance. Under this condition the evaporating atoms make negligibly few collisions with the atoms of the residual gas in the chamber. The following table gives the values of the mean free path of an atom of silver at various pressures of residual oxygen.

7

Table 2.1

Pressure mm Hg	Mean free path cm
10^{-3}	4·5
10^{-4}	45
10^{-5}	450
10^{-6}	4500
10^{-7}	45000

For a typical laboratory system with dimensions of the order of tens of centimetres, the mean free path condition is met by working at a pressure of 10^{-5} mm Hg. There is, however, an additional and more stringent condition to be fulfilled if the influence of the residual gas on the film structure is to be reduced to negligible proportions. This concerns the rate at which residual gas atoms strike the surface of the target during the deposition of the film. On kinetic theory it is seen that the number of gas molecules at pressure p and absolute temperature T striking unit area of a plane surface in unit time is given by

$$N = p/(2\pi mkT)^{\frac{1}{2}} \qquad \qquad .... \ 2(1)$$

For oxygen at 10^{-5} mm Hg and at room temperature the value of N is $5\cdot2 \times 10^{15}$ atoms cm^{-2} sec^{-1}. Taking the area occupied by a single adsorbed atom of oxygen on the surface to be $\sim 1\cdot4$ Å^2 it is seen that at this pressure the rate at which the residual atoms strike the target is such as to form a monomolecular layer in ~ 1 second. Since this rate of deposition is not small compared with that for the condensing material under the usual evaporation conditions, it is clear that, if interference by the residual gas is to be avoided, work must be carried out at pressures very much lower than 10^{-5} mm Hg. In this connection, it may be mentioned that, in their work on the electrical properties of thin films of the alkali metals, E. T. S. APPLEYARD and A. C. B. LOVELL[1] found it necessary to bake the target surface for 80 hours at $250°$ C at a pressure below 10^{-7} mm Hg before reproducible results could be obtained.

A further example is provided by the behaviour of films of aluminium prepared in a vacuum system employing a mercury diffusion pump. Such films are attacked by mercury, resulting in a breaking-up of the film. This is observed even when trapping with solid carbon dioxide and acetone is used although the vapour pressure of mercury at this temperature ($-78°$ C) is only 3×10^{-9} mm Hg.

It is thus clear that for any work on the fundamental aspects of the condensation process a glass or silica vacuum system, preferably without greased cone joints, which may be subjected to prolonged baking, is a necessity.

Effect of beam intensity

The influence of beam intensity is shown in a striking manner by the experiments of Wood, Knudsen and others on cadmium, silver, and mercury. The existence was shown of a critical beam density below which no condensation occurred and also of a critical substrate temperature, above which no film could be formed except at very high beam intensities. These experiments and their interpretation are discussed in Section 2.6 below.

The target surface

The effects of the target surface on the nature of the condensed film depend on the structure of the substrate, whether amorphous, polycrystalline or monocrystalline, on the target temperature, and on the cleanliness of the target. The studies of the epitaxy of condensed layers show clearly the influence of the first factor; the condensing atoms tend to take on the structure type of the underlying surface, forming amorphous layers on amorphous substrates and single-crystal layers on single-crystal substrates. This behaviour is itself markedly temperature-dependent. Films of germanium are found to be practically amorphous when deposited on a monocrystalline surface at room temperature. On raising the temperature, the film becomes polycrystalline and, at a higher temperature, monocrystalline (H. König[2], L. E. Collins and O. S. Heavens[3]). Unless precautions are taken to maintain the target surface at a steady temperature during the deposition of a film, the adventitious rise of temperature of the surface during the evaporation may result in a film possessing any of a wide range of structure types.

In observations of the structure of antimony deposited (a) on a freshly-formed surface of gold and (b) on a surface contaminated by a film of grease, S. Levinstein[4] found that, in the former, the film formed was continuous (within the resolution limit of some 100Å) whilst on the contaminated surface, aggregates of diameters ranging up to 1–2 μ were formed.

Influence of rate of evaporation

Variation of the rate of evaporation has the effect of varying both the intensity of the condensing beam and also the mean velocity of arrival of the atoms. Separation of these two factors is a matter of some difficulty although, since the range of temperature over which evaporation is practicable tends to be rather limited, the available range of velocities of arrival is not large. By the use of a velocity selector, Levinstein investigated the effect of varying the beam velocity on the structure of films of gold, bismuth and antimony and found, for velocities within thermal ranges, no detectable influence on structure. This result, although negative, is of considerable value in indicating that variations in structure observed on varying the rate of evaporation result mainly from the variation of beam intensity produced. The difficult experimental problem of separating the two factors is thus removed.

Levinstein observed that low rates of evaporation of these metals tend to produce non-crystalline layers, or layers in which the crystal size is very small ($\sim$10–15 Å) whereas high rates of evaporation yield films of large crystallites. Similar behaviour has been reported for chromium[5].

Contamination by the source material

In much of the work which has been carried out on evaporated films, the method of evaporation using a boat or filament as the supporting source (see Section 2.3) has been employed. Tungsten, tantalum or molybdenum are frequently used for this purpose, since they possess high melting-points. It is known that, in certain instances, e.g. the evaporation of aluminium from a tungsten spiral, solution of the supporting material by the molten charge may occur. The resulting solid solution usually loses the charge material preferentially during the evaporation so that the film formed contains a much smaller proportion of the spiral material than does the molten charge. The extent of the contamination of germanium and silver films by these source materials has been determined by O. S. HEAVENS[6] using radioactive tracer techniques and has been found not to exceed a few parts per million under the best conditions. It is unlikely that an impurity content of this order will influence the optical properties of films, except in the case of semiconducting materials in the infra-red region of the spectrum. If this degree of contamination is to be avoided, other techniques of evaporation employing high frequency heating may be employed.

An exhaustive study of the influence of all the factors enumerated above on each substance to be examined would be a time-absorbing and unrewarding process. There are indications that certain of the above factors have negligibly small effect on certain properties of films. Thus Levinstein found no detectable differences in the *structures* of films deposited at different pressures of residual gas over a range extending up to 10^{-3} mm Hg. The *electrical* properties of films of germanium have, however, been found to be significantly different for films deposited at 10^{-4} mm Hg from those formed at lower pressures. The results of P. L. CLEGG[7] on the optical properties of silver films show little variation of the optical behaviour with rate of deposition.

2.3. PRACTICAL METHODS OF EVAPORATION

The oven

As mentioned above, for work on the fundamental aspects of the condensation process, an all-glass vacuum system, in which pressures of less than 10^{-8} mm Hg are obtainable, is the first prerequisite. The material to be evaporated must be contained in an oven so that accurate control of the temperature of the source may be effected. Since under the normal evaporation conditions the mean free path of the evaporating atoms is large compared with the dimensions of the system, and therefore with the size of the aperture of the oven, conditions of molecular streaming obtain. The intensity, I, of the beam of atoms leaving the oven in a direction making an angle θ with the normal to the orifice, is given by the cosine law.

If p = vapour pressure of the material in the oven
$\quad T$ = absolute temperature
$\quad A$ = area of orifice
$\quad r$ = distance of target from the oven, in the direction θ then

$$I = \frac{N_0}{(2\pi MRT)^{\frac{1}{2}}} \cdot \frac{pA}{\pi r^2} \cos \theta \qquad \text{.... 2(2)}$$

where N_0 is Avogadro's number, R the gas constant and M the molecular weight of the substance evaporating.

On a plane surface (*Figure 2.1*) at a distance y from the oven, the number of atoms arriving per unit area of the surface in a direction θ is given by

$$I_1 = \frac{N_0}{(2\pi MRT)^{\frac{1}{2}}} \cdot \frac{pA}{\pi y^2} \cos^4 \theta \qquad \text{.... 2(3)}$$

The oven is heated electrically and the temperature measured near the orifice, using a thermocouple. The relations 2(2) and 2(3) apply only under the mean free path condition stated. If the mean free path is comparable with the slit dimensions, then a cloud of vapour forms near the slit, evaporation takes place effectively from this cloud and the distribution obtained on the condensing surface is characteristic of an extended source. For an oven made from metal, the problem of obtaining an even temperature distribution is not generally difficult. There are, however, many substances

11

for which no suitable metal exists for use as a container of the molten charge. Thus although aluminium may be evaporated from a tungsten helix, it cannot satisfactorily be melted in bulk in a tungsten oven, on account of the rapid solution of the tungsten in the aluminium. For such materials, a refractory lining, e.g. of graphite, porcelain, quartz, thoria, etc., is required.

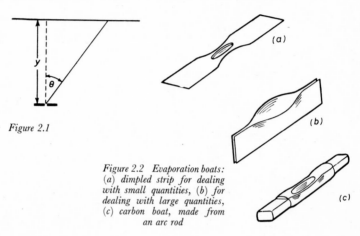

Figure 2.1

Figure 2.2 Evaporation boats: (a) dimpled strip for dealing with small quantities, (b) for dealing with large quantities, (c) carbon boat, made from an arc rod

For the preparation of films under conditions where the temperature of the source need not be known with high accuracy, or for the evaporation of materials for which the use of an oven is quite impracticable, there are several simpler types of source, described below.

The strip or boat

This is perhaps the simplest of all methods of evaporation and may be employed for a wide variety of materials. A strip, or shaped boat, of high melting-point material is heated by passing a large current, the material to be evaporated being placed on the strip or in the boat. Typical forms of boat are shown in *Figure 2.2*. Tungsten, molybdenum and tantalum in thicknesses of about 0·1– 0·2 mm are frequently used as strip materials and appear to be suitable for almost all dielectric materials (except the most refractory) and for many metals. With certain metals, particularly aluminium, solid solution of these boat materials by the molten evaporation charge is known to occur and this has the effect of weakening the boat. However, if the boat is mounted in a reasonably strain-free position, fracture may be avoided. The difficulties

12

occasioned by solution of the boat material may often be avoided by the use of a carbon boat which can conveniently be made from an arc rod.

Preliminary heating of the boat to drive off surface contamination is necessary; the oxides of tungsten, molybdenum and tantalum

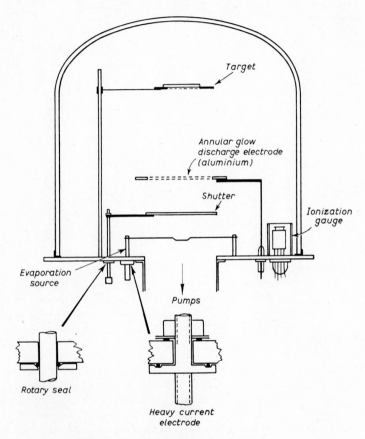

Figure 2.3   Layout of typical evaporation plant

volatilize readily at temperatures well below the melting-points of the metals.   When the evaporation is started, a shutter is interposed between source and target in order to intercept the first fraction of the evaporate, which latter may contain impurities from the surface of the charge.   In mounting metal strips or boats, care should be taken to avoid materials (e.g. brass screws) which may volatilize low melting-point constituents on heating. A typical demountable vacuum system is shown in *Figure 2.3*. Rubber O ring seals may be used for sealing electrodes and for rotating shafts. A wide glass tube serves to shield the ionization gauge from the evaporation source. The glow discharge electrode is made of aluminium, for which the rate of sputtering is very low. The glow discharge supply should be of a few kilovolts and 10–20 mA (for a 40 cm diameter chamber) and the supply for the evaporation source of order 5–50 V at 200–20 A. For further details of the experimental arrangements, reference [8] may be consulted.

By only a slight modification of the above system, the evaporation of alloys may be effected. The boat is heated empty to a temperature well above that at which the constituents of the alloy evaporate rapidly. The powdered alloy is then dropped in a slow stream into the boat. L. HARRIS and B. M. SIEGEL[9] have examined the composition of films of α-brass, β-brass and of a gold–cadmium alloy produced by this method and have found the composition to be the same as that of the alloy before evaporation to within close limits.

The helical coil

Where solid solution of the boat material occurs so rapidly as to render the strip or boat heater unduly liable to fracture, the use of a helical coil of wire is frequently adopted. Wire of 0·5–1·0 mm diameter is coiled into a helix of about 5 mm diameter, and with turns so spaced that they are not short-circuited by the material being evaporated. This procedure has been used by J. STRONG[10] for the evaporation of aluminium, a tungsten helix being used. The aluminium is attached in the form of small loops of wire. (There would appear to be no limit to the scale on which this method may be applied since this was the method used for aluminizing the 200-inch mirror for the Mount Palomar telescope.) Evaporation takes place mainly from the sides of the filament, rather than from the blobs of aluminium. Surface tension causes the aluminium to run up from the blob and, since the sides of the helix are at a higher temperature than the more massive blobs, the rate of evaporation from this region is higher. According to Strong, tungsten and aluminium form a solid solution containing up to 3 per cent tungsten. The solution loses aluminium preferentially, however, so that the evaporate is substantially free from tungsten.

An alternative method of loading the helix is to electroplate it with the material to be evaporated. This has been successfully employed for the evaporation of iron, nickel and rhodium.

The conical basket

For metals which do not wet the material of the wire, or for those for which the vapour pressure below the melting-point is high enough ($\sim 10^{-2}$ mm Hg) to produce a sufficiently high rate of evaporation, this method is simple and effective. The supporting wire is wound round the last three turns or so of an ordinary woodscrew. For brittle metals such as tungsten and molybdenum, this is done in a bunsen flame. Chips of the material to be evaporated are then dropped into the basket so formed. This method has been used for chromium and cobalt.

When a metal is to be evaporated which wets these supporting materials, the short-circuiting of the turns may be prevented by the use of a small alumina or beryllia crucible with the spiral wound round it. It is difficult to obtain high temperatures by this method without radiating considerable heat to the target surface. As an alternative, the spiral may be coated with a paste of aluminium oxide and sodium silicate and the spiral preheated to yield an adherent, insulating layer on the turns of the wire. A list of elements which have been successfully evaporated by these last two methods has been given by L. O. OLSEN, C. S. SMITH and E. C. CRITTENDEN [11].

Although experimentally very simple and convenient, the types of source listed above possess certain disadvantages. Accurate control of temperature, and hence rate of evaporation, is not easy. With the boat type of heater, this difficulty may be minimized by the use of a large boat which is kept well filled. If only a small fraction of the charge is evaporated, the change in resistance, and hence the tendency for the temperature to rise, may be kept small. A thermocouple may conveniently be spot-welded on the underside of the centre of the boat, or a radiation pyrometer used for the temperature measurement. With the helix or spiral types of heater there appears to be no simple way of overcoming this difficulty.

Electron bombardment

For the evaporation of highly refractory materials, the method of electron bombardment has been applied by H. M. O'BRYAN [12]. The charge is contained in a graphite crucible which is surrounded by a tungsten helix (*Figure 2.4*). The helix is heated to 2300° C and is maintained at -4kV to the crucible, the whole being surrounded by a radiation shield of tantalum. Very high temperatures are attainable by this method, the evaporation of carbon (at 3500° C) having been effected. Thorough outgassing of the crucible is essential since the electron bombardment method requires high vacuum conditions in order to be effective.

15

Induction heating

The method of induction heating has been applied to the evaporation of metals in order to remove the possibility of contamination of the films by the material supporting the source. A globule of metal to be evaporated is placed on a refractory rod and surrounded by a water-cooled coil which is fed from a supply at a frequency of a few kilocycles per second.

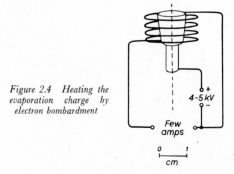

Figure 2.4   Heating the evaporation   charge   by electron bombardment

2.4. THE METHOD OF SPUTTERING

Before the development of the thermal evaporation techniques described above, the method of sputtering was widely used for the preparation of films. For certain metals, e.g. platinum and molybdenum, for which the thermal evaporation method is difficult on account of their high melting-points, the method of sputtering still forms a convenient way of obtaining films. The process involves maintaining a discharge in a gas, preferably inert, at a pressure of about 10^{-3} mm Hg, the cathode being made of the metal to be sputtered. According to A. VON HIPPEL[13] the mechanism is essentially one of local boiling of the cathode surface resulting from intense local heating arising from bombardment by the positive ions in the discharge. It would be expected, in this case, that metals which are normally covered with a refractory oxide layer (e.g. aluminium) would exhibit a very low rate of sputtering and this is found to be so. The experiments of R. W. DITCHBURN[14] show that the behaviour of sputtered particles in the vapour phase is in every way similar to that of particles produced by boiling. The absence of a critical beam density and deposition temperature, observed in the deposition of films by the evaporation method, is ascribed to the removal of adsorbed gas on the target surface by

16

the positive ions of the gas discharge. Condensation would then occur on the unmasked target surface. Recently a sputtering technique has been evolved which enables this process to be used with much lower gas pressures than hitherto. A strong magnetic field is maintained between the cathode surface and the target. Electrons emitted by the cathode then make spiral paths and travel a much longer distance before reaching the target than would be the case in the absence of the field. The spiralling electrons will thus make many more collisions with gas molecules at a given gas pressure, or alternatively will make the same number of collisions at a much lower pressure, than will electrons in the absence of the field.

Little work has been done in recent years either on the theoretical aspects of cathodic deposition or on the properties of sputtered films. From the nature of the deposition process, such films must be expected to contain occluded gas and are on this account less suitable for study than are films formed under high vacuum conditions.

2.5. OTHER METHODS OF FILM DEPOSITION

Electrolytic deposition

Although films of many metals may be easily formed by the method of electrolysis, little attention has been paid to the study of the properties of such films. The method suffers from the disadvantage that the number of factors on which the properties of the films may be expected to depend is large and that many of these factors are not easy to control. The fact that the films are produced in the presence of an electrolyte makes it likely that they will contain adsorbed foreign molecules. The structure of the deposits depends on the purity of the electrolyte used, and on the current density. The results of examination of the structures of electrodeposited films are given in Section 3.9.

Chemical and other methods

As with electrolytically prepared films, little attention has been paid, in so far as their optical properties are concerned, to films deposited by purely chemical means. Banning has used films of titanium dioxide deposited from the vapour of titanium tetrachloride as highly-reflecting layers (Section 7.3), the light absorption in such layers being very much smaller than that of a transparent metal film having the same reflectivity.

Anti-reflecting films on glass surfaces have been formed by bombarding the surface with high-speed (60 keV) ions of krypton, argon and xenon. The bombardment produces a layer on the surface of lower refractive index than that of the glass itself and for which the refractive index varies continuously in a direction normal to the surface. This type of film, discussed in more detail in Chapter 7, is of importance since it enables the reflectivity of a surface to be reduced over a greater range of wavelength than does a film of constant refractive index.

2.6. FUNDAMENTAL ASPECTS OF THE CONDENSATION PROCESS

Early work

The early experiments of Wood (1915) showed that, when a stream of metallic vapour impinged on a surface, the rate of condensation depended on the surface temperature. Working with mercury vapour, he deduced that all the incident atoms condensed at temperatures below − 140° C. Knudsen (1916) showed that, for zinc, cadmium and mercury, a critical temperature existed above which no condensation occurred and found this critical temperature to lie between − 183° C and − 78° C. The work of Chariton and Semenoff (1924) and Estermann (1925) showed that the critical temperature depended on the intensity of the impinging beam of atoms. References to early work in this field are given in a paper by Ditchburn[14].

The results of these experiments were adequately described by Frenkel in the following way.

Frenkel theory of adsorption

The failure of the impinging atoms to condense on a surface whose temperature is many hundred degrees below the boiling-point of the metal indicates that the atoms cannot be adsorbed by the substrate surface immediately on arrival. The reason for this may be the presence of an adsorbed gas layer on the surface, which prevents the condensing atom from coming within reach of the true surface forces of the substrate. Frenkel postulated that the atoms are free to move over the surface on arrival and do so for a certain mean free time τ after which they re-evaporate. They may, however, collide with other metal atoms while on the surface, and so form pairs. The mean free time of such pairs is shown to be greater than that of a single atom, so that, with the formation of pairs, the chance of further collisions increases and the building-up of nuclei

18

and hence of a thick film is facilitated. The existence of a critical surface temperature and critical beam intensity follow from these assumptions, as is shown below.

If ν is the number of gas atoms striking a surface S in unit time and n the number of atoms on the surface when equilibrium is established, then the mean time τ for which a condensed atom remains on the surface is given by

$$\tau = n/\nu \qquad \text{.... } 2(4)$$

If the number of atoms in the vapour phase which are in equilibrium with the n adsorbed atoms is denoted by n' and if $\bar{v}$ is the mean value of the component of velocity of these atoms perpendicular to the surface S, then

$$\nu = (S/2V)n'\bar{v} = (Sn'/V)(kT/2\pi m)^{\frac{1}{2}} \qquad \text{.... } 2(5)$$

by the use of Maxwell's law of distribution of velocities, where k is Boltzmann's constant, T the absolute temperature and m the mass of the atom.

Use is made of the result from statistical mechanics that the number of particles of energy W whose co-ordinates (position and velocity) occupy a volume $d\omega$ of phase-space is given by $\exp(-W/kT)d\omega$. Integration over a region of phase-space gives the number of particles covered by the volume concerned, the integral being termed the reduced phase-volume. Thus if Φ is the reduced phase-volume for the adsorbed particles and Φ' that for the particles in the gaseous phase, we have

$$n/n' = \Phi/\Phi' \qquad \text{.... } 2(6)$$

To calculate Φ', we assume that the atom may oscillate with period τ_0 normally to the surface S, about a position at which its potential energy is $-u_0$. The change in potential energy, Δu, due to a displacement z from the mean position, is given by

$$\Delta u = (2\pi^2/\tau_0^2)mz^2 \qquad \text{.... } 2(7)$$

and the total potential energy in the displaced position is $(-u_0 + \Delta u)$. Since the condensed and solid phases are in equilibrium, the mean velocities of the particles in the two phases are equal so that

$$\Phi/\Phi' = (1/V)\int_{-\infty}^{\infty} S \exp\{(u_0 - \Delta u)/kT\}dz \qquad \text{.... } 2(8)$$

Substituting from 2(6) and 2(7) we obtain

$$n/n' = (S\tau_0/V)(kT/2\pi m)^{\frac{1}{2}}\exp(u_0/kT) \qquad \text{.... } 2(9)$$

19

and from 2(5)

$$\tau = \tau_0 \exp(u_0/kT) \qquad \text{.... } 2(10)$$

In the foregoing treatment it has been assumed that there is no interaction between atoms on the surface, so that u_0 is the energy of a single, isolated atom on the surface. In close proximity to another atom, the energy of one of the atoms on the surface becomes $u_0 + \Delta u_1$ and the corresponding mean free time becomes

$$\tau_1 = \tau_0 \exp\{(u_0 + \Delta u_1)/kT\} = \tau' \exp(u_0/kT) \qquad \text{.... } 2(11)$$

where $\tau' = \tau_0 \exp(\Delta u_1/kT)$. The mean free time of a pair of atoms thus exceeds that of a single atom. If the atom under consideration is surrounded by r other atoms at the same distance, then the energy becomes $u_0 + r\Delta u_1$, and the group has a correspondingly longer mean free time. The reciprocal of τ may be interpreted as the probability for re-evaporation to occur; this probability is thus lower for aggregates of atoms than for single atoms.

If σ_0 is the area on the surface occupied by a single atom, i.e. the area over which the field of force of the atom is appreciable, then when a total of n atoms occupy a surface of area S (and provided that n is small compared with the number required to form a monatomic layer) the number of atoms which lie within the fields of force of other atoms is given by $n(n-1)\sigma_0/S$*. The effective surface area of a pair of atoms is given by

$$\sigma = \sigma_0 \exp(\Delta u_1/kT)$$

so that the number of paired atoms is

$$n_1 = n(n-1)\sigma/S \fallingdotseq n^2\sigma/S (\text{since } n \gg 1) \qquad \text{.... } 2(12)$$

The probability that a single atom re-evaporates is given by $1/\tau$ and that a pair evaporate is $1/\tau'$. The rate of change of the number of atoms on the surface is thus given by

$$\mathrm{d}n/\mathrm{d}t = \nu - n_1/\tau' - (n - n_1)/\tau \qquad \text{.... } 2(13)$$

since there are, at time t, n_1 pairs and $(n - n_1)$ unpaired atoms on the surface.

Putting $\alpha = 1/\tau$ and $\beta = (\sigma/S)(1/\tau - 1/\tau')$ into equation 2(13), we see that if $\nu < \alpha^2/4\beta$, then n increases to the limiting value $\alpha/2\beta$

---

* For, if a single atom lies on the surface and $(n-1)$ are added, the number falling within the field of force σ_0 of the first atom is $(n-1)\sigma_0/S$. Since there are n atoms altogether, the total number of such overlaps is $n(n-1)\sigma_0/S$.

whereas if $v > a^2/4\beta$, $dn/dt > 0$ and the number of atoms in the layer steadily increases. The theory thus accounts for the existence, at a given temperature, of a critical beam density below which a thick film may not be built up. The dependence of β on temperature, through the σ term, shows the existence of a critical temperature, for a given beam density, above which no condensation will occur.

The dependence of critical beam density on temperature is readily shown to be given by

$$v_c = [4\sigma_0\tau_0 \exp(u_0 + \Delta u_1)/kT]^{-1} \qquad \text{.... } 2(14)$$

where it is assumed that $\tau' \gg \tau$. The energy of binding of an atom pair will be expected to be much greater than kT, so that this approximation is justified.

The validity of the Frenkel hypothesis

The results of Cockcroft, on the condensation of cadmium on a copper surface give considerable support to the assumptions of the Frenkel theory.

A polished copper disc was placed so as to receive the stream of cadmium vapour from an oven. The temperature of the oven, and hence the stream intensity, was varied as was also the temperature of the receiving surface. The critical density phenomenon was shown by the formation on the copper disc of a circular patch of cadmium which grew to a definite maximum diameter. Beyond the limits of this circle no deposition occurred, however long the experiment was continued. Since, from the cosine law of distribution of the atoms, the density of the stream falls steadily from the centre of the system, it is clear that the density beyond the patch observed is too low for condensation to occur.

The relation between critical density and temperature of the target was determined over the temperature range $-155°$ C to $-80°$ C, over which the critical density ranged from 4×10^{12} to 2×10^{16} atoms cm^{-2} sec^{-1}. A linear relation between log (critical density) and 1/temperature, shown in *Figure 2.5*, was observed, as is predicted by the Frenkel theory [equation 2(14)].

The results of Estermann indicate that a deposit some 3–4 atoms thick is visible. The Frenkel theory enables the time taken to form a visible deposit to be calculated and hence the form of the law of growth of the circular patch of metal in the Cockcroft experiment to be calculated. The predicted form is found to agree with that observed. It may be further shown that for stream densities which exceed the critical density by a factor of four, the effect of re-evaporation of the condensed atoms may be neglected.

The observations described above leave little doubt as to the essential correctness of the Frenkel hypothesis. The observations on surface motion by Estermann, and by R. W. Ditchburn[14] and K. I. ROULSTON[15] lend further support to the theory. That the mobility of the atoms over the surface is a consequence of the presence of adsorbed gas on all 'natural' surfaces is made very probable by

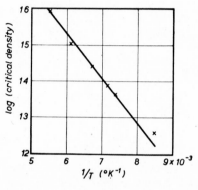

Figure 2.5

Cockcroft's experiments in which cadmium was condensed on a freshly deposited silver surface, formed by evaporation. Much lower critical densities were observed than with natural surfaces, even when the latter had been heated in a vacuum to 900° C. The results of Ditchburn, mentioned below, on sputtered films also point to this general conclusion.

Work on sputtered films

Working with cadmium, deposited by the method of sputtering, Ditchburn showed that the critical phenomena observed with evaporated films do not occur. The results of the experiments, in which the potential of the target surface was varied so that positive ions of various speeds were collected, suggest that the effect of the ion bombardment is to free the surface from adsorbed gas films so that direct adsorption of the condensate may occur.

References

[1] APPLEYARD, E. T. S. and LOVELL, A. C. B.   *Proc. Roy. Soc.* **A 158** (1937) 718

[2] KÖNIG, H.   *Reichsber. f. Phys.* **1** (1944) 4

[3] COLLINS, L. E. and HEAVENS, O. S.   *Proc. Phys. Soc.* **B 65** (1952) 825

[4] LEVINSTEIN, S.   *J. Appl. Phys.* **20** (1949) 306

[5] COLLINS, L. E. and HEAVENS, O. S.   *J. de Phys.* **13** (1952) 658

[6] HEAVENS, O. S.   *Proc. Phys. Soc.* **B 65** (1952) 788

[7] CLEGG, P. L.   *Ibid.* 774

[8] HEAVENS, O. S.   *Lab. Practice* **1** (1952) 157

[9] HARRIS, L. and SIEGEL, B. M.   *J. Appl. Phys.* **19** (1948) 739

[10] STRONG, J.   *Modern Physical Laboratory Practice.*   Blackie, London, 1940

[11] OLSEN, L. O., SMITH, C. S. and CRITTENDEN, E. C.   *J. Appl. Phys.* **16** (1945) 425

[12] O'BRYAN, H. M.   *Rev. Sci. Inst.* **5** (1934) 125

[13] VON HIPPEL, A.   *Ann d. Physik* **80** (1926) 672; **81** (1926) 1043

[14] DITCHBURN, R. W.   *Proc. Roy. Soc.* **A 141** (1933) 169

[15] ROULSTON, K. I.   *Proc. Camb. Phil. Soc.* **37** (1941) 440

THE STRUCTURE OF SOLID FILMS

General review of methods of investigation—Results of optical methods of examination—The method of electron diffraction—The electron microscope—Results of electron-optical studies—Influence of the electron beam—The sensitization phenomenon in film deposition—Field emission microscopy—The structure of electrodeposited films

3.1. GENERAL REVIEW OF METHODS OF INVESTIGATION

FOR a complete understanding of the behaviour of a thin film, a knowledge of the structure of the film is necessary. From the information which has been gathered by the methods discussed below, it is apparent that only in rare cases does a film consist of a homogeneous parallel-sided layer. Films are found to exhibit a wide variety of structures, from irregular, amorphous aggregates to monocrystalline layers. The properties of such layers, and especially the optical and electrical properties, would be expected to depend considerably on the exact form of the film.

Most of our present knowledge of the behaviour and structure of thin films has come to us by one or more of three methods of examination. A limited amount of information is obtainable by direct (visual) optical microscopy, whilst detail on a smaller scale is afforded by the use of electron-optical methods, viz. by electron diffraction and by electron microscopy. In methods employing visible light microscopy, the limited resolving power enables only gross features, such as large-scale aggregation of crystallites in the film, to be examined.

The electron-optical methods enable information on the detailed structure of the films to be obtained. The electron microscope, which gives a direct image of the object, enables a resolution limit of some 10 Å to be attained, thus bringing within range crystallites which contain only a few tens of atoms. The information obtainable by electron diffraction is of a less direct nature than that given by microscopy, but enables details of the crystal structure of the material in the film to be obtained on a truly atomic scale. It is also possible to follow chemical changes in the film from the changes in the electron diffraction pattern. Thus although no recognizable changes are observed in the electron micrograph of a metal film when a small amount of oxidation occurs, the presence of oxide is readily shown by the electron diffraction pattern.

The visible-optical method is in a sense complementary to the electron-optical methods, each providing information within its own scale. There is at present a gap between the region served by electron diffraction and that for which electron microscopy is suitable. Work with the field emission microscope (Section 3.8) indicates that this method may be able to yield information in this region, although the range of investigations for which this method is applicable is somewhat limited.

3.2. RESULTS OF OPTICAL METHODS OF EXAMINATION

Among the earliest work on the process of film formation by condensation from the vapour is that of K. ESTERMANN[1] on the deposition of silver and cadmium. It was found that when condensation of a beam of atoms of uniform cross-section occurred, the distribution of material in the films was microscopically non-uniform. The deposits were found to consist of groups of isolated nuclei, indicating that lateral motion of the atoms had taken place after deposition.

Figure 3.1 Pattern obtained between crossed polarizers from a sputtered silver film

The existence of this surface motion was further confirmed by J. COCKCROFT[2] using vaporized cadmium and was also exhibited by sputtered cadmium (K. I. ROULSTON[3]). E. N. DA C. ANDRADE and J. G. MARTINDALE[4] examined the effect of heating on films of silver and gold deposited on amorphous substrates by sputtering and were able to observe the growth of crystallites in the films by optical microscopy. The films were examined between crossed polarizers, patterns of the type shown in *Figure 3.1* being obtained. These patterns are characteristic of spherulitic aggregates of uniaxial crystallites. It is suggested that the crystallites take the form of long needle-like crystals of the order ten atoms across and that anisotropy arises from the limitation in extension normal to the needle axis. On theoretical grounds (J. E. LENNARD JONES[5]) it is expected that the lattice spacing normal to the needle axis would differ from that along this axis. Direct confirmation of this variation in spacing is difficult on account of the inherent broadening of diffraction rings from such small crystallites. It is significant, however, that the spherulitic form is obtained only for small crystallites; heating to higher temperatures (350° C) results in the formation of large crystallites, seated on (111) faces, a result which is confirmed by the electron diffraction work of K. R. DIXIT[6]. [Orientations other than (111) have been observed in such films.]

These experiments illustrate the type of information which may be obtained by visual optical methods and afford further confirmation of the high degree of mobility possessed by atoms in thin layers even at temperatures several hundred degrees below the melting-point.

3.3. THE METHOD OF ELECTRON DIFFRACTION

The general features of the crystal structure of a thin film may be readily obtained from an examination by electron diffraction. If the film contains only minute crystallites, the diffraction pattern consists of very diffuse rings (*Figure 3.2*). The rings occur even with an amorphous film since the interatomic distances are in this case distributed about a mean value. The mean spacing between atoms in a truly amorphous layer differs from the lattice spacing of the crystal. The radii of the diffuse rings obtained with most evaporated films generally indicate the presence of minute crystallites, rather than an amorphous layer. The presence of large crystalline regions in the film is revealed by the occurrence of sharp rings in the pattern (*Figure 3.3*), the continuity of which indicates that there is no preferred orientation of the crystals in the film. The same pattern will thus be obtained for all orientations of the film with respect to the beam. It is possible for continuous rings to appear even if the film consists of oriented crystallites, this condition occurring when the beam is parallel to the fibre axis. On tilting the specimen, however, the rings break up into arcs. An estimate of the size of crystallites in the film may be made from the radial breadth of the diffracted beams (G. P. THOMSON and W. COCHRANE[7]). Close inspection of the rings in the pattern from a

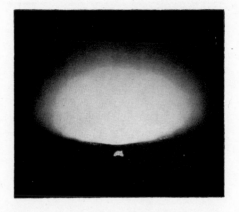

Figure 3.2   Electron diffraction pattern of amorphous chromium film

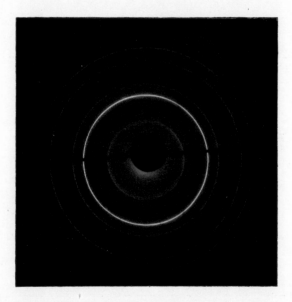

Figure 3.3   Electron diffraction pattern of polycrystalline gold. Continuous rings are obtained at all angles of incidence indicating random orientation of crystallites

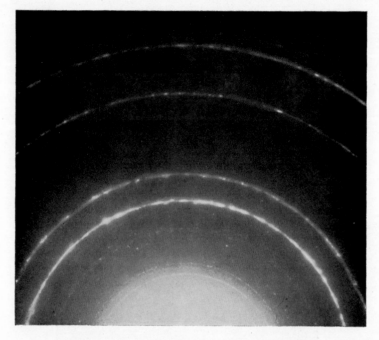

Figure 3.4   Spots in polycrystalline pattern from diffraction by individual crystallites

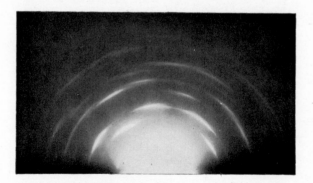

Figure 3.5   Reflection electron diffraction pattern of aluminium showing (111) fibre orientation

polycrystalline specimen reveals spots from individual crystallites, as shown in *Figure 3.4.*

When the crystallites are lying so that one of their crystallographic axes is parallel, or nearly parallel, to a fixed direction, but with the other axes directed at random, fibre orientation results, yielding a diffraction pattern similar to that shown in *Figure 3.5.* This behaviour is often observed in films deposited by the evaporation method and occurs even when deposition takes place on an

Figure 3.6   Monocrystalline film of nickel, formed by condensation on heated rocksalt

amorphous (e.g. glass) substrate. The direction of the fibre axis is found sometimes to depend on the direction of evaporation. The degree of fibre orientation may be found from the length of the arcs in the pattern.

Condensation on a heated monocrystalline substrate may yield

29

films which are themselves monocrystalline, the process being termed epitaxy. Thus *Figure 3.6* was obtained from a thin film of nickel deposited on a rocksalt (100) face, maintained at 450° C. Since epitaxy is observed only with monocrystalline substrates, it is clear that the periodicity associated with the single-crystal nature of the substrate is an important factor governing this phenomenon. A close identity between the lattice spacings of the condensing film and those of the substrate is apparently unnecessary, however, as is illustrated by the results of D. W. PASHLEY[8]. The conditions for epitaxy to occur have been studied extensively; the occurrence of epitaxial growth and the types of orientation which result are found to depend both on the closeness of the lattice spacings and on the coordination across the boundary between deposit and substrate. The relative importance of these two factors has been discussed by L. E. COLLINS[9] for several metals.

Apart from information on the general structure of films, outlined above, the diffraction method may enable detailed information on crystal shape to be obtained. Thus when cobalt is deposited on a monocrystalline substrate, doubling and quadrupling of the diffraction spots is observed, indicating the development of octa-hedral faces of the condensed material. The existence of twinning in crystalline layers is also readily deduced from the occurrence of additional spots in the diffraction pattern. Additional spots may also occur when the film is mechanically strained.

When the radii of the rings of a diffraction pattern from a poly-crystalline evaporated film are measured with high accuracy, anomalies are found which may be interpreted as arising from strains in the crystallites resulting from the existence of surface tension forces. (T. B. RYMER and C. C. BUTLER[10]).

The existence of surface motion of the condensing atoms (Section 2.6) is further confirmed by electron diffraction observations on very thin layers. [Metals, L. H. GERMER[11]; alkali halides, L. G. SCHULZ[12].] When a quantity of material is condensed which would, if evenly distributed, form a layer some 10 Å in thickness, the breadth of the diffraction rings obtained shows the presence of crystallites some 50 Å across. The range of movement across the surface must, as pointed out by Schulz, be extremely limited, partly on account of the uniformity of the observed crystal size and partly from the dependence of the direction of the fibre axis in oriented deposits on the direction of evaporation. Any considerable degree of mobility of the atoms after condensation would cause the influence of the direction of arrival to be masked.

3.4. THE ELECTRON MICROSCOPE

The very high resolving power which can be attained by the use of the electron beam instead of visible light waves makes the electron microscope a particularly valuable instrument for the study of thin films. By the use of this instrument information is obtained directly on the structures of thin films, whereas hitherto such information could be obtained only indirectly from studies of optical and electrical properties. In many cases, the structures proposed in order to account for the optical and electrical behaviour of films have been substantially confirmed by electron micrographs of the films.

Development of the electron microscope has proceeded rapidly in the last two decades. At the time of writing, the resolving power of the microscope is in the neighbourhood of 10 Å. This is considerably above the theoretical Abbe resolution limit for the wavelengths which are used ($\lambda \sim 0.01$ Å) and is limited by the aberrations in the electron lens systems employed which restricts the maximum usable aperture of the microscope to a value much lower than that attainable in a microscope using visible light. Thus whilst crystallites or aggregates containing a few tens of atoms may be observed, the single atom is still not in sight so far as the electron microscope is concerned. On account of the low penetrating power of the electron beams used in electron diffraction and in electron microscopy, their use is limited to films whose thickness does not exceed a few thousand ångström units.

General features of the structures of evaporated films

In the preceding chapter, mention was made of the considerable evidence for the mobility of atoms over a substrate surface and of the resultant tendency for aggregates to form. Another mechanism which tends to favour an aggregated structure is to be found in the large surface tension forces which are known to exist in solids. The pressure exerted on a drop from the action of surface tension varies inversely as the radius of the drop. The presence of any irregularity or discontinuity in a continuous thin film therefore favours the breaking-up of the film into droplets. An extreme case of droplet formation is shown in *Figure 3.7* of a tin film of average thickness 10,000 Å.

The optical behaviour observed in certain evaporated films lends further support to the idea of an aggregated rather than a continuous structure. Transparent films of zinc and cadmium deposited at the temperature of liquid air are found to possess highly reflecting surfaces. On warming to room temperature, the

31

transmittance of the films is found to increase. The reflectance decreases and the films show considerably greater scatter than is shown at the low temperature. This behaviour is consistent with the presence of considerable aggregation at the higher temperature resulting in large interstices between crystallites. The scattering suggests a crystallite size which is not small compared with the light

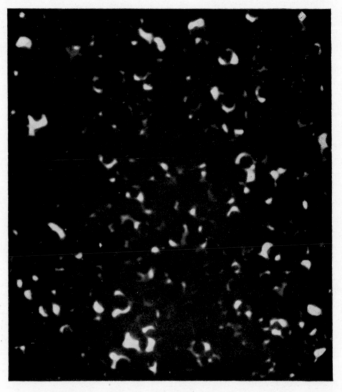

Figure 3.7   Aggregation into droplets in a tin film (× 250)

wavelength used. Still further evidence for a discontinuous structure springs from measurements of the conductivity of thin metal films. Below a limiting thickness, the conductivity is practically zero, suggesting the presence of small crystallites which are sufficiently isolated to afford no continuous conducting path along the film. The electron microscope has enabled direct confirmation to be obtained of the granular structure inferred from the experiments described above.

3.5. RESULTS OF ELECTRON-OPTICAL STUDIES

(i) *Alkali halides*

Using both electron diffraction and electron microscopy, Schulz[12] has examined the structures of films of several alkali halides deposited on glass. The optical properties of these films in the visible have also been studied and the results are considered in more detail in Section 5.6. Lithium fluoride and sodium chloride have been most extensively studied; substantially similar general behaviour has been observed in films of KCl and CaF_2. For film thicknesses below about 100 Å, random orientation of crystallites or a slight tendency to (100) orientation is obtained. As the film thickness increases well-marked fibre orientation develops. For LiF and CaF_2, (111) orientation occurs. For NaCl and KCl (110) and for MgF_2 (302) orientations are observed. On account of the tendency of LiF and NaCl to grow by the development of (100) planes it would be expected that films showing a considerable degree of (111) orientation would exhibit a pyramidal structure, showing (100) facets. This was observed directly by Schulz in electron micrographs of thick films, the micrographs being obtained by the replica technique. Thicknesses of LiF up to about 70,000 Å were examined. Owing to the pronounced development of pyramidal crystallites, which in thick films may be many hundreds of ångströms across at the base, such films scatter visible light considerably and show a pronounced white bloom.

The general conclusions from these experiments, in which the films were deposited on substrates of collodion or glass, are that the growth of dielectric films is not influenced by the substrate when this is amorphous, and that the film consists of loosely-fitting particles with appreciable amounts of empty space round them. Thus even for films of LiF 18 μ thick, solid LiF amounted to only 82 per cent of the total film bulk. The growth of crystallites in this way is readily understood from the known mobility of the atoms on arrival at the surface. The uniformity observed in the sizes of crystallites shows, however, that the range of motion of the atoms across the surface cannot exceed a few hundred ångström units. The results of studies on metal films which we now discuss leads to a similar view.

(ii) *Metal films*

H. LEVINSTEIN[13] has examined the structures of evaporated films of a large number of metals using both electron diffraction and electron microscopy. The films were formed by evaporation on a collodion at a pressure not exceeding 10^{-5} mm Hg. Films of several different thicknesses were examined and it was found to be possible to classify the electron diffraction results in three groups, according to the type of pattern obtained.

(a) *Diffuse rings.*—This group consists almost entirely of metals with melting-points above 1900° C. The diffuse patterns are found to indicate a crystalline, rather than an amorphous, state, with a crystal size less than about 15 Å. No appreciable variation with film thickness was observed for metals in this group. The metals falling into this group are: germanium, iridium, niobium, rhodium, silicon, tantalum and tungsten.

(b) *Sharp rings. No preferred orientation.*—The melting-points of the metals in this group mainly fall between 600° C and 1900° C. For very thin films, diffuse rings are obtained; with increasing film thickness, the rings become sharp. The metals in this group are: beryllium, chromium, cobalt, copper, gold, iron, lead, manganese, nickel, palladium, platinum, silver, tin and titanium.

(c) *Sharp rings. Preferred orientation.*—This is the low melting-point group, consisting exclusively of metals with melting-points below 650° C. There is generally a tendency for the crystallites to lie with the most closely-packed plane parallel to the substrate. The metals exhibiting this type of pattern are: antimony, bismuth, cadmium, indium, magnesium, tellurium, thallium and zinc.

The above classification is intended only as a broad indication of the differences in structures for different melting-points. The structures of metals condensed on glass surfaces frequently differ from those of films formed on collodion. Thus thick gold films are found to show marked preferred orientation when condensed on glass although the melting-point of gold is above 650° C. Germanium films may form with any of the three structure types mentioned above, depending on the substrate temperature. Moreover, it would seem difficult to avoid heating (with consequent softening of) the collodion in depositing thick layers of the high melting-point metals. The combination of elevated surface temperature and instability of substrate would certainly influence the process of crystallization of the films. However, within these limitations, the classification gives a fair picture of the relation between the film structure and the melting-point of the metal.

The electron microscope was used by Levinstein to study the effect of varying certain of the conditions of deposition on the type of structure obtained. The results of these experiments emphasize the need for care in using the instrument in that the electron beam was found markedly to affect certain of the films examined. This is further discussed in Section 3.6 below.

The pressure of the residual gas in the vacuum chamber was found to have no effect on the structures of the films for pressures up to 10^{-2} mm Hg for materials which do not react with the residual gas. The rate of evaporation was found to have a marked effect, especially for metals with a low melting-point. With antimony, low rates of deposition ($\sim$ one atom layer per second) were found to give films consisting of isolated droplets; the electron-diffraction patterns were very diffuse. Rapid evaporation produced films with a considerable degree of aggregation of the droplets. These deposits were crystalline. In the slowly-evaporated films the droplets were about 100–200 Å across. That the droplets observed in the electron micrographs cannot be individual crystallites is seen from the diffuse diffraction patterns obtained. Crystallites of this size would yield sharp rings.

The use of a velocity selector enabled the effect on the film structure of variation of the velocity of arrival of the atoms to be examined. For metals whose vapour is monatomic, no variation was found within thermal ranges. For certain metals (e.g. antimony), the vapour was found to be polyatomic and for these materials, the grain size in the film varied with the size of the molecules in the vapour.

Films of aluminium, cadmium, copper, gold, magnesium and zinc have been studied in the electron microscope by R. G. PICARD and O. S. DUFFENDACK [14]. The relative sizes of the particles and the interstices between them were noted. Differences were sometimes observed in the structures obtaining when different substrates were used and the following classification was introduced to characterize the different structure types.

A. Surfaces which crystallize with increasing temperature to form agglomerates separated by interstices which are much narrower than the diameters of the aggregates.

B. Surfaces which crystallize with increasing temperature to form agglomerates separated by interstices of the same order of magnitude as the particle diameters.

C. Surfaces which recrystallize with increasing temperature to form relatively isolated crystals.

The results of applying this scheme of classification to the metals examined are collected in *Table 3.1*.

Table 3.1

Metal	Substrate	Class
Aluminium . . .	Collodion	A
	Glass	A
Cadmium . . .	Collodion	C
	Glass	C
	Copper	B
Copper . . .	Collodion	A
	Glass	A
Gold	Collodion	A
	Glass	A
Magnesium . . .	Collodion	A
	Glass	A
Zinc	Collodion	C
	Glass	C
	Aluminium	B
	Copper	B

These results may be compared with a classification introduced by E. T. S. APPLEYARD[15] based on the theory of adsorption developed by Lennard Jones[5]. The substrate surface is characterized by the existence of potential troughs over the substrate atoms, the troughs being separated by barriers which may or may not impede the movement of atoms across the surface, depending on their kinetic energy (E) and on the latent heat of evaporation (L_m) of the metal from itself. If L_s is the latent heat of evaporation of a metal atom from the substrate and if $L_s > L_m$, then no lateral surface motion will occur so that a uniform monatomic film would be stable. Such films cannot be observed with the electron microscope. If $L_s < L_m$ and if $L_s \gtrsim 1$ electron volt, then at low temperatures, coherent, stable films could be formed. At high temperatures, the larger kinetic energy of the atoms on the surface will result in lateral motion with resultant agglomeration. If $L_s < L_m$ and L_s is only $\sim 0 \cdot 1$ eV, then surface motion can occur even at liquid-air temperature. The kinetic energy required for atoms to surmount the low potential barriers on the surface is so small that aggregation occurs even at very low surface temperatures. The last-mentioned class is seen to correspond with Picard and Duffendack's class C. Films which are classified under A and B fall under the second of Appleyard's categories, with $L_s < L_m$ and $L_s \gtrsim 1$ eV.

The electron microscope has thus provided direct evidence of the granular structure of thin films of solids and in doing so has enabled many of the properties of films to be understood. The attainable

36

resolution is not quite sufficient to deal with very thin layers of certain materials, which appear structureless by this mode of examination although other methods of investigation show them to be porous. The so-called 'anomalous' optical and electrical properties of thin films are anomalous only on the assumption that the film is a parallel-sided continuum. The electron microscope has shown this not to be so and has provided a true, albeit depressing, picture of a thin film. For, although the granular structure of the films enables their properties to be appreciated in a qualitative way, the nature of the structures is such as to make quantitative description a matter of extreme difficulty.

3.6. INFLUENCE OF THE ELECTRON BEAM

In the examination of materials by electron diffraction or electron microscopy, the possible influence of the electron beam on the film structure must not be overlooked. From the nature of the process it is clear that some of the energy of the incident electrons is transferred to the specimen and that some heating will occur. The lack of quantitative data in the various reports indicating the presence or absence of influence by the beam makes it impossible to specify the conditions under which specimens may be safely examined.

Since the current density at the specimen in the electron microscope is very much higher than that usually used in the electron diffraction camera, it is in the microscope that trouble from this cause is more likely. Levinstein[13] has shown, with what is described as a 'strong' electron beam, that films of thallium melted and then volatilized. Zinc and tellurium were found to sublime without melting, whilst antimony and gold underwent a process of crystallization. The experiments of R. S. SENNETT and G. D. SCOTT[16] show that films of silver are unaffected by the electron beam in the microscope and that they are aggregated before examination by the electron beam. This is shown by shadow-casting the film before examination; the aggregates observed in the micrograph are found to possess shadows. The writer has found that the diffraction pattern due to stretched cellulose (which yields a typical fibre pattern) disappears within 2–3 seconds when examined in an electron beam of density 20 μA/cm^2 due, presumably, to melting of the specimen. G. I. FINCH[17] has, however, obtained quite sharp crystalline patterns from paraffin wax (melting-point 45° C) showing that the temperature rise under the conditions of his examination cannot have exceeded a few degrees. Furthermore, in the precision determination of the lattice parameters of caesium iodide, T. B. RYMER and P. G. HAMBLING[18] were able to estimate from the expansion coefficient of this material that the temperature rise was less than 10° C. These results are of considerable importance in showing that diffraction patterns *can* be obtained under conditions such that the rise in specimen temperature is no more than a few degrees. The results of Sennett and Scott show equally that electron micrographs *can* be obtained without disruption of a silver specimen. Those of Levinstein indicate clearly the need for circumspection in dealing with materials of low melting-points. Whilst reduction of the beam intensity and consequent lengthening of exposure may be expected to reduce the danger of specimen heating, this procedure introduces further difficulties, such as that of the formation of a film of cracked hydrocarbon, or ultimately carbon itself, on a specimen exposed to prolonged electron bombardment in a chamber evacuated by an oil diffusion pump. Besides the effect on the structure of the film itself, the presence of the electron beam has been observed to inhibit markedly the formation of a germanium film from the vapour. (L. E. COLLINS and O. S. HEAVENS[19].)

An approximate value for the rise in temperature of a specimen of thickness d, irradiated area A and of material in which the mean free path of electrons is λ gives

$$\text{Rise in temperature} = 2 \times 10^3 \, iVd/\lambda A \quad (^\circ\text{C})$$

$$\text{where } i = \text{beam current (amps)}$$

$$\text{and } V = \text{accelerating potential (volts).}$$

This indicates that if the temperature rise is to be less than 20° C in a specimen 50 Å thick in which the mean free path is 1000 Å in a beam accelerated by 50,000 volts, then the current density at the specimen should not exceed 4 μA/cm². Since the current densities which are usually found in electron diffraction cameras range through the value obtained above, it is to be expected that discrepant accounts of the influence of the beam will arise when low melting-point materials are examined.

3.7. THE SENSITIZATION PHENOMENON IN FILM DEPOSITION

In the last chapter, the mechanics of film formation were discussed and were shown to be well described by the Frenkel theory. When a surface is exposed to a beam of atoms of intensity below the critical value, and at a temperature above the critical value, then re-evaporation from the surface occurs at such a rate as to prevent the formation of a thick film. The surface is then partly covered by a mobile, two-dimensional layer from which re-evaporation takes place in equilibrium with the condensation of the impinging atoms. In order that a thick film shall form, the temperature of the surface and the rate of arrival of atoms must be such that pairs or small aggregates can form, whose longer mean free time on the surface enables larger and larger nuclei to form.

Early experiments by E. N. DA C. ANDRADE and H. TSIEN [20] on sodium, and by Ditchburn on cadmium indicate the possibility of film formation at beam intensities below the critical value and at temperatures above the critical value if there exist suitable nuclei on the surface. Andrade observed that sodium vapour will deposit in minute surface fissures which are found on drawn glass tubing although no deposition occurs on the smooth parts of the surface. Ditchburn showed that condensation of cadmium does not occur on such a surface unless the surface has been thoroughly outgassed. The mechanism of condensation in such fissures is clearly that the atoms condensing have only restricted mobility, along the direction of the crack, so that the probability of the formation of nuclei is considerably enhanced over that for a flat surface. On a surface which has not been outgassed, the presence of gas atoms in the fissures presumably prevents the entry of the cadmium atoms. Further confirmation of the suggested mechanism is provided by the result that no such condensation occurs on a blown surface.

Sensitization

The experiments discussed above show that the presence of nuclei adsorbed on the surface is necessary in order that a thick film may be built up. For a complete description of the growth process, it is necessary to know what surface concentration of nuclei is required in order that a thick film may be formed. This quantity is related to, and may be estimated from, the range of the atoms on the surface. In experiments by Ditchburn, a glass surface was exposed to a weak beam of cadmium vapour, the beam intensity being determined from the known values of vapour pressure. The intensity was such that 10^{-3} of a monatomic layer would deposit in 5 seconds. The target was first cooled by liquid air. The results of Estermann referred to in Section 2.6 show that under this condition all the incident cadmium atoms are adsorbed. It is thus possible to estimate the number of adsorbed cadmium atoms down to very small fractions of a monatomic layer. On allowing the surface, on which a measured small fraction of a layer of cadmium atoms had been thus deposited, to attain room temperature, it was found that a visible deposit of cadmium formed within a small number of minutes. On the unsensitized surface and with the same beam intensity, no deposit formed in $1\frac{1}{2}$ hours. Some typical results of these experiments are given in *Table 3.2*. Under the conditions

Table 3.2. *Dependence on Quantity of Sensitizing Material of Time to form Visible Deposit*

Quantity of sensitizing material (fraction of monolayer)	Time to form visible deposit (secs)	Target condition
0·001	17	not outgassed
0·004	$4\frac{1}{4}$	
0·013	6	
0·025	7	
0·0014	14	
0·0030	15	
0·0049	12	
0·016	$3\frac{1}{4}$	outgassed
0·035	3	
0·30	$3\frac{1}{4}$	
0·90	3	

of the experiment, a visible deposit formed on the liquid-air cooled target surface in $1\frac{1}{4}$ minutes. In the absence of the sensitizing deposit and with the target at room temperature, no deposit formed in $1\frac{1}{2}$ hours. The results show clearly the profound effect on film formation of amounts of adsorbed cadmium corresponding to ∼0·001 molecular layer.

In view of the observed ranges of atoms in surface motion experiments the above results are not surprising. The effective range on the surface over which a nucleus may collect migrating atoms is such as to cover a very large number of atoms. F. M. FOLEY[21] examined deposits of cadmium a few atoms thick and found nuclei in the deposits surrounded by circular regions free from deposit, showing that migration into the nucleus had taken place. Diameters of the clear patches up to ∼0·05 mm were observed. *Figure 3.8* shows a similar effect observed in a tin film.

Sensitization studies using electron microscopy

The structures of films of zinc and cadmium have been examined using electron microscopy by E. ZEHENDER[22]. The structures of these films, and also their electrical conductivity, are found to be markedly influenced by the presence of a sensitizing layer. Silver was used as a sensitizing material in these experiments, in amounts from 10^{-3} to 10^{-1} monolayer. The beam intensities used in the zinc and cadmium deposition were considerably above the critical value

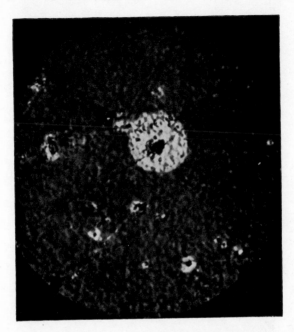

Figure 3.8   Migration of atoms to form a nucleus (× 230)

so that a thick film could be built up at room temperature even in the absence of a sensitizing deposit.

Typical micrographs obtained with different amounts of sensitization are shown in *Figure 3.9*. The presence of 10^{-2} sensitizing monolayer results in a smooth deposit of cadmium (or zinc) which has the lustre characteristic of the polished bulk metal and whose electrical conductivity is 70–80 per cent of that of the bulk metal, even for films ~ 200 Å thick. Electron diffraction patterns show very sharp rings from deposits sensitized with a 10^{-2} monolayer, in contrast to the diffuse rings and dense background observed with the films sensitized with a 10^{-3} monolayer. The latter films appear blue by reflected light and show a bloom characteristic of aggregated deposits. The blue colour results from Tyndall scattering by the crystallites. Films deposited on an unsensitized substrate appear grey and show an even greater degree of aggregation. They are practically non-conducting and show no adhesion to the substrate.

Figure 3.9 Influence of small amounts of silver on the form of condensed cadmium (a) 10^{-2} monolayer of Ag ; (b) 10^{-3} monolayer ; (c) no Ag deposit

3.8. FIELD EMISSION MICROSCOPY

In the field emission microscope, electrons are withdrawn from the specimen under examination by the application of very high (~millions of volts/cm) fields at the surface of the specimen. The specimen is conveniently a very fine point at the centre of a spherical flask, the inner surface of which is coated with a semi-conducting fluorescent powder, which latter surface acts as anode. The attainable magnification depends on the ratio of the screen radius to the radius of curvature of the specimen point. By suitable etching techniques, the latter may be made as small as ~1 micron so that, with a 10 cm radius screen, a magnification of ~10^6 is attainable. Magnifications of this order have been obtained with this instrument (E. W. MÜLLER[23]). The limit of resolution is governed by the component of the electron velocity at right angles to the radius joining the specimen to the screen and corresponds to a resolution limit of about 2 Å. Whilst not quite sufficient to enable individual

42

atoms to be resolved, the method has enabled images of fairly large molecules (e.g. copper phthalocyanine) to be distinguished. The images resemble closely the form of the molecules deduced from X-ray crystallographic studies.

The field emission microscope has been used so far to study the adsorption of various metals on crystalline faces of tungsten and molybdenum. Much useful information has been obtained enabling the adsorption process to be described in terms of the crystal structures of the adsorbed materials and those of the adsorbent. When a layer of foreign atoms is adsorbed on a crystal face, the work function changes so that a change in intensity of the image of that face is observed. For adsorption to occur it would be expected that the interatomic spacings of the adsorbate would need to be fairly close to those of the atoms in the substrate crystal planes so that the adsorbed layer is substantially free from strain. From a consideration of the atomic spacings in the substrate crystal planes, we may deduce on which planes a given foreign atom layer would be expected to be adsorbed. Such predictions are substantially confirmed by many workers. An excellent review of the work in this field is given by F. Ashworth [24].

3.9. The Structure of Electrodeposited Films

We conclude this chapter with a description of the results of investigations on the structures of metal films deposited electrolytically. Such films have not proved of great interest from the standpoint of their optical properties (except perhaps as reflectors for car headlamps). Their study is of interest, however, in connection with the mechanism of film growth. The conditions of film formation in an electrolytic bath are somewhat more complicated than those attending the growth of films deposited by thermal evaporation.

Both optical microscopy and electron diffraction have been used to study these films and the influence of the bath conditions (temperature, current density, bath composition) on the structure of the substrate are reasonably well understood. G. I. Finch and D. N. Layton [25] distinguish two types of crystal growth. In one of them growth occurs outwardly from nuclei on the substrate, with negligible deposition between such nuclei, whilst in the second type growth takes place laterally so that deposition occurs over the whole of the substrate surface. The former type of growth is favoured by (i) high current density, (ii) low ion concentration, (iii) a low bath temperature and (iv) a low circulation rate of the electrolyte. These conditions all tend to produce local depletion of the strength of the electrolyte on the surface in the neighbourhood of a growing nucleus so that growth occurs outwardly rather than laterally. Conditions which tend to maintain high ion concentration over the whole surface favour lateral growth over the surface and result in smooth deposits contouring the substrate. Thus whilst a current density of 0·1 mA/cm² in an N/2 silver nitrate bath results in a deposit which is smooth to within 0·01 mm, rapid deposition with a current density of 2000 mA/cm² produces flowery growths extending a millimetre or more from the surface.

Studies of the structures and orientations of electrodeposits by electron diffraction have shown clearly how the mode of growth is influenced by the substrate structure. When deposition of a metal occurs on a single crystal of the same metal under conditions such that lateral growth takes place, the electrodeposit is found to continue the lattice of the substrate. The single crystal nature of electrodeposited iron on a (110) face of an iron crystal has been observed at thicknesses up to 30,000 Å and would presumably continue thus indefinitely. Similar behaviour has been observed on (100), (310), (321) and (332) faces of iron crystals and also on crystals of copper, tin and bismuth. This type of growth is also observed in films produced by thermal evaporation on to monocrystalline substrates; epitaxial growth occurs only if the substrate temperature is raised to the point at which the atoms have sufficient mobility to migrate to the potential troughs over the atoms of the substrate. In electrodeposition, the electrolyte provides the necessary mobility for the cations and epitaxy occurs at room temperature, provided lateral growth conditions are established.

When electrodeposition takes place on a polycrystalline substrate, the structure of the initial layers is determined by that of the substrate. At large thicknesses, the crystal structure is determined by the bath conditions, the substrate influence being lost. The smaller the crystal size of the substrate, the more easily is the deposit structure influenced by the bath conditions. This would be expected since the smaller the crystal size, the larger is the proportion of intercrystalline boundary regions, at which the substrate influence is small. In these regions, the crystal growth is thus much more strongly influenced by the bath conditions.

REFERENCES

[1] ESTERMANN, K.   *Z. Phys. Chem.* **106** (1923) 403; *Z. f. Phys.* **33** (1925) 320

[2] COCKCROFT, J.   *Proc. Roy. Soc.* **A 119** (1928) 293

[3] ROULSTON, K. I.   *Proc. Camb. Phil. Soc.* **37** (1941) 440

[4] ANDRADE, E. N. DA C. and MARTINDALE, J. G.   *Phil. Trans.* **235** (1935) 69

[5] LENNARD JONES, J. E. and DENT, B. M.   *Proc. Roy. Soc.* **A 121** (1928) 247

[6] DIXIT, K. R.   *Phys. Zeits.* **39** (1938) 580

[7] THOMSON, G. P. and COCHRANE, W.   *Theory and Practice of Electron Diffraction.* Macmillan, London, 1939.

[8] PASHLEY, D. W. *Proc. Roy. Soc.* **A 210** (1952) 354

[9] COLLINS, L. E.   Ph.D. Thesis, University of Reading, 1954

[10] RYMER, T. B. and BUTLER, C. C.   *Proc. Phys. Soc.* **59** (1947) 541

[11] GERMER, L. H.   *Phys. Rev.* **56** (1939) 58

[12] SCHULZ, L. G.   *J. Chem. Phys.* **17** (1949) 1153; *Act. Cryst.* **4** (1951) 483

[13] LEVINSTEIN, H.   *J. Appl. Phys.* **20** (1949) 306

[14] PICARD, R. G. and DUFFENDACK, O. S.   *Ibid.* **14** (1943) 291

[15] APPLEYARD, E. T. S.   *Proc. Phys. Soc.* **49** (1937) E118

[16] SENNETT, R. S. and SCOTT, G. D.   *J.O.S.A.* **40** (1950) 203

[17] FINCH, G. I.   *Trans. Far. Soc.* **31**-II (1935) 1144

[18] RYMER, T. B. and HAMBLING, P. G.   *Act. Cryst.* **4** (1951) 565

[19] COLLINS, L. E. and HEAVENS, O. S.   *Proc. Phys. Soc.* **B 65** (1952) 825

[20] ANDRADE, E. N. DA C. and TSIEN, H.   *Proc. Roy. Soc.* **A 159** (1937) 346

[21] FOLEY, F. M.   M.Sc. Thesis, Trinity College Dublin, 1941

[22] ZEHENDER, E.   *Optik* **7** (1950) 200

[23] MÜLLER, E. W.   *Phys. Zeits.* **37** (1936) 838

[24] ASHWORTH, F.   *Advances in Electronics*, Vol. III, 1951

[25] FINCH, G. I. and LAYTON, D. N.   *J. Electrodepos. Tech. Soc.* **27** (1951) Adv. Copy No. 9

4

THIN FILM OPTICS

The problem of notation—Reflection and transmission of light at the surface of a transparent medium—Reflection at the surface of an absorbing medium—Reflection and transmission of light by a single film—Reflection and transmission of light by two films—The extension to a system of multiple layers—Optical impedance—Matrix method using Fresnel coefficients—Application of matrix method for evaluating reflectance and transmittance—Graphical methods (approximate for transparent layers)—Graphical methods (general)—The phase change on reflection at or transmission through a thin film—General theory of doubly refracting films

4.1. THE PROBLEM OF NOTATION

IN principle, the determination of the amplitudes and intensities of beams of light reflected or transmitted by a thin film or by a system of several films is straightforward. We simply set up Maxwell's equations and apply the appropriate boundary conditions. In practice, the resulting equations are depressingly complicated and the evaluation of the properties of a given combination of films involves monstrously tedious computation. The problem has been dealt with in several ways, some of which are described below. Several different approaches are used and, broadly speaking, for each approach a different notation has been employed. This does not, as might first appear, signify any rooted objection among different workers to the establishment of a unified notation but has arisen partly from the fact that many of the different treatments were evolved at about the same time and partly because certain notations better suit one type of approach than do others. The difficulty in finding a suitable notation is appreciated when it is realized that when, for example, the amplitude of an electric vector is specified, it is necessary to include in the symbol the direction of propagation and the direction of polarization. Furthermore, an indication of the medium in which the wave is travelling is also required, particularly when, as in the case of multiple layers, there may be several media with distinct properties in the system considered. The symbol for the wave amplitude must therefore show these three quantities in addition to announcing itself as an electric vector. Precisely similar considerations apply to the magnetic vector.

The notation adopted in this book for wave vectors is a slight modification of that used by F. ABELÈS[1]. So far as indicating the direction of propagation is concerned, it is sufficient if the wave symbol shows whether the wave travels in the positive or negative sense with respect to some fixed direction (usually the normal to the film or system of films). A superscript + or − therefore serves this purpose. In specifying the direction of the plane of polarization, it is sufficient to indicate whether this is parallel to or perpendicular to the plane of incidence. For, in dealing with problems involving light polarized in a plane of arbitrary orientation with respect to the plane of incidence, the invariable procedure is to resolve the wave vectors into components in and normal to the plane of incidence. In multilayer problems it is convenient to specify the number of the layer by a simple suffix. Our complete specification, then, for the amplitude of the electric vector of a wave travelling in the positive direction in the nth layer, or medium, in the system and polarized with the electric vector parallel to the plane of incidence is E_{np}^{+}. We use E_{ns}^{+} for the component of the electric vector perpendicular to the plane of incidence.

For the case of normal incidence on an *isotropic* medium, it is unnecessary to show the direction of polarization so that we denote the electric vector of the positive-going wave in the nth medium by E_{n}^{+} and that of the negative-going wave by E_{n}^{-}. For the corresponding magnetic vectors, we use H_{n}^{+} and H_{n}^{-}. The use of Fresnel coefficients in writing expressions for reflected and transmitted light enables this simplified notation to be used sometimes even for non-normal incidence (see Section 4.5).

Absorbing media

The equations of propagation of light in a transparent medium may be used to describe propagation in an absorbing medium if the refractive index n is replaced by a complex quantity, of which the imaginary part is related to the absorption of energy by the medium. For a plane wave entering the plane boundary of an isotropic, absorbing medium, the planes of equal phase of the wave in the medium are perpendicular to the direction of propagation. Since the reduction in amplitude of the wave in the medium depends directly on the distance travelled in the medium, the loci of points of equal amplitude will be planes parallel to the surface of separation.

Only for the case of normal incidence are the planes of equal phase parallel to those of equal amplitude.

Thus, if we consider a wave of circular frequency ω travelling in the direction (λ, μ, ν) in a transparent medium of refractive index n, the electric vector may be written

$$E = E_0 \exp i\omega\left\{ t - \frac{n(\lambda x + \mu y + \nu z)}{c} \right\} \qquad \dots 4(1)$$

where c is the velocity of light in vacuum.

The corresponding expression in an absorbing medium becomes

$$E = E_0 \exp i\omega\left\{ t - \frac{a(\lambda x + \mu y + \nu z)}{c} + \frac{i\beta(\lambda' x + \mu' y + \nu' z)}{c} \right\} \qquad \dots 4(2)$$

where (λ', μ', ν') is the direction of maximum damping—i.e. the direction of the normal to the planes of equal amplitude.

For normal incidence, the expression for the wave may be written

$$E = E_0 \exp i\omega\left\{ t - \frac{(n - ik)(\lambda x + \mu y + \nu z)}{c} \right\} \qquad \dots 4(3)$$

since, for this case, the direction of maximum damping coincides with the direction of propagation.

In equation 4(3), n is the ratio of the velocity of the wave in vacuum to that of the wave in the medium. k represents the energy absorption: the attenuation of the amplitude of the wave for a path of one *vacuum* wavelength in the medium is $\exp(-2\pi k)$. It must be emphasized that this applies only to a wave for which the planes of equal phase are parallel to those of equal amplitude. In equation 4(2), the values of a and β depend on the direction of propagation in the medium and hence on the angle of incidence. If the angle of incidence is θ and the angle between the planes of constant phase and those of constant amplitude is φ, then it follows directly from the wave equation* that

$$a^2 - \beta^2 = n^2 - k^2 \qquad \dots 4(4)$$

$$a\beta \cos \varphi = nk \qquad \dots 4(5)$$

$$\sin \theta = a \sin \varphi \qquad \dots 4(6)$$

The fact that the equation of propagation of a wave entering an absorbing medium normally may be expressed in similar form to

---

* See, e.g., Ditchburn's *Light* (Blackie), p. 480.

that for a transparent medium by replacing the (real) index n by a complex quantity has resulted in $(n-ik)$ being termed the 'complex refractive index'. The name is unfortunate for two reasons. (1) Refractive index is formally defined as the ratio of two velocities and is therefore of necessity real and (2) the real part of the so-called complex index does not bear the relation to the angles of incidence and refraction which is borne by the value n for the transparent medium. The relation between θ, φ and n may be obtained from equations 4(4) to 4(6) and is much more involved than Snell's law.

More serious, however, than the adoption of an unsuitable name for the index is the existence of two distinct (although not distinct enough) notations, viz:

$$\boldsymbol{n} = n - ik \qquad \text{.... } 4(7)$$

and
$$\boldsymbol{n} = n(1 - \varkappa) \qquad \text{.... } 4(8)$$

with the further confusion that 4(8) is sometimes written $\boldsymbol{n} = n(1 - ik)$. As mentioned earlier, k as defined by equation 4(7) represents the attenuation of the wave per *vacuum* wavelength. $\varkappa$ is related in a precisely similar way to the attenuation per wavelength in the medium for the wave transmitted at normal incidence.

As the widespread use of both notations indicates, there is no logical preference for the one over the other. Neither is in any sense the more fundamental. The form $\boldsymbol{n} = n - ik$ will be used throughout this book; the resulting equations are faintly less cumbersome.

4.2. REFLECTION AND TRANSMISSION OF LIGHT AT THE SURFACE OF A TRANSPARENT MEDIUM

For an isotropic medium, the laws of electromagnetism are represented by the following relationships

$$\text{div } \mathbf{D} = \varepsilon \text{ div } \mathbf{E} = 4\pi\varrho \qquad \text{.... } 4(9)$$

$$\text{div } \mathbf{D} = \mu \text{ div } \mathbf{H} = 0 \qquad \text{.... } 4(10)$$

$$\text{curl } \mathbf{E} = -\frac{\mu}{c}\frac{\partial \mathbf{H}}{\partial t} \qquad \text{.... } 4(11)$$

$$\text{curl } \mathbf{H} = \frac{4\pi\sigma\mathbf{E}}{c} + \frac{\varepsilon}{c}\frac{\partial \mathbf{E}}{\partial t} \qquad \text{.... } 4(12)$$

where the symbols have their usual meanings. Electrical quantities are measured in electrostatic units and magnetic quantities in electromagnetic units. For a medium in which there is no space-charge, these relations lead directly to Maxwell's equations, representing the propagation of electromagnetic disturbances in the medium.

$$\frac{\varepsilon\mu}{c^2}\frac{\partial^2\mathbf{E}}{\partial t^2} + \frac{4\pi\mu\sigma}{c^2}\frac{\partial\mathbf{E}}{\partial t} = \nabla^2\mathbf{E} \qquad \dots 4(13)$$

$$\frac{\varepsilon\mu}{c^2}\frac{\partial^2\mathbf{H}}{\partial t^2} + \frac{4\pi\mu\sigma}{c^2}\frac{\partial\mathbf{H}}{\partial t} = \nabla^2\mathbf{H} \qquad \dots 4(14)$$

For propagation in a non-conducting ($\sigma = 0$) medium, these reduce to

$$\frac{\varepsilon\mu}{c^2}\frac{\partial^2\mathbf{E}}{\partial t^2} = \nabla^2\mathbf{E} \qquad \dots 4(15)$$

$$\frac{\varepsilon\mu}{c^2}\frac{\partial^2\mathbf{H}}{\partial t^2} = \nabla^2\mathbf{H} \qquad \dots 4(16)$$

the well-known simple form of wave equation, which shows that disturbances are propagated with velocity $c/\sqrt{(\mu\varepsilon)}$. Since, at optical frequencies the value of μ for all materials is insensibly different from unity, the velocity of propagation is $c/\sqrt{\varepsilon}$ where ε is the dielectric constant at the frequency of the light wave. From the definition of refractive index, we have the well-known result $n = \sqrt{\varepsilon}$.

The problem of determining the light reflected and transmitted at a boundary separating two media is dealt with by applying boundary conditions to the solutions of Maxwell's equations. These require that the tangential components of both electric and magnetic vectors be continuous at the boundary, which latter is for the moment envisaged as a mathematically sharp discontinuity separating regions of differing optical properties. We consider only sinusoidal solutions of the equations. Solutions for other types of wave may in principle be dealt with by the aid of Fourier's theorem. Any variation of the propagation parameters with wavelength must, of course, be taken into account in effecting the summation.

We consider a plane wave incident on the surface $z = 0$, the plane of incidence being the plane $x0z$, the angle of incidence φ_0 and the angle of refraction φ_1. We assume that the surface is infinite in extent so that it may accommodate the necessarily infinitely wide plane wave of unique frequency which we are supposing to be incident upon it. (From the practical point of view this limitation is unimportant since the errors involved for the smallest experimentally practicable beams is negligible.)

The co-ordinate system is shown in *Figure 4.1*. In accordance with the notation discussed in Section 4.1, we denote the amplitudes of the electric vectors of the wave approaching the surface by E^+_{op}

and E_{os}^+ for the two components. The reflected wave is denoted by E_{op}^-, E_{os}^- and the transmitted wave by E_{1p}^+, E_{1s}^+. The phase factors associated with the incident and reflected waves are of the form

$$\exp i\left(\omega t - \frac{2\pi n_0 x \sin \varphi_0}{\lambda} - \frac{2\pi n_0 z \cos \varphi_0}{\lambda}\right) \quad \text{(incident)}$$

and

$$\exp i\left(\omega t - \frac{2\pi n_0 x \sin \varphi_0}{\lambda} + \frac{2\pi n_0 z \cos \varphi_0}{\lambda}\right) \quad \text{(reflected)}$$

whilst that for the transmitted wave is

$$\exp i\left(\omega t - \frac{2\pi n_1 x \sin \varphi_1}{\lambda} - \frac{2\pi n_1 z \cos \varphi_1}{\lambda}\right)$$

where λ is the wavelength in vacuum.

At the boundary, which we take at $z = 0$, the point of incidence being the origin of co-ordinates, we have for the total components of the electric and magnetic vectors in the x- and y- directions

$$\left.\begin{aligned}
E_{0x} &= (E_{0p}^+ + E_{0p}^-) \cos \varphi_0 \\
E_{0y} &= E_{0s}^+ + E_{0s}^- \\
H_{0x} &= n_0(-E_{0s}^+ + E_{0s}^-) \cos \varphi_0 \\
H_{0y} &= n_0(E_{0p}^+ - E_{0p}^-)
\end{aligned}\right\} \quad \dots\ 4(17)$$

for the first medium and

$$\left.\begin{aligned}
E_{1x} &= E_{1p}^+ \cos \varphi_1 \\
E_{1y} &= E_{1s}^+ \\
H_{1x} &= -n_1 E_{1s}^+ \cos \varphi_1 \\
H_{1y} &= n_1 E_{1p}^+
\end{aligned}\right\} \quad \dots\ 4(18)$$

Applying the boundary conditions, we obtain equations which may be solved to give the amplitudes of the transmitted and reflected vectors in terms of those of the incident vectors. We obtain

$$\frac{E_{0p}^-}{E_{0p}^+} = \frac{n_0 \cos \varphi_1 - n_1 \cos \varphi_0}{n_0 \cos \varphi_1 + n_1 \cos \varphi_0} = r_{1p} \qquad \dots\ 4(19)$$

$$\frac{E_{1p}^+}{E_{0p}^+} = \frac{2n_0 \cos \varphi_0}{n_0 \cos \varphi_1 + n_1 \cos \varphi_0} = t_{1p} \qquad \dots\ 4(20)$$

$$\frac{E_{0s}^-}{E_{0s}^+} = \frac{n_0 \cos \varphi_0 - n_1 \cos \varphi_1}{n_0 \cos \varphi_0 + n_1 \cos \varphi_1} = r_{1s} \qquad \dots\ 4(21)$$

$$\frac{E_{1s}^+}{E_{0s}^+} = \frac{2n_0 \cos \varphi_0}{n_0 \cos \varphi_0 + n_1 \cos \varphi_1} = t_{1s} \qquad \dots\ 4(22)$$

51

r_{1p}, r_{1s} are known as the Fresnel reflection coefficients and t_{1p}, t_{1s} the Fresnel transmission coefficients. We shall later see how the solutions for the case of multiple layers are conveniently expressed in terms of these coefficients.

From equations 4(19)–4(22) we see that $t_{1p} = 1 + r_{1p}$ and $t_{1s} = 1 + r_{1s}$ so that, for the case $n_0 > n_1$, the values of t_{1p} and t_{1s} exceed unity. This may at first sight appear strange since the coefficients have been defined as the ratios of the amplitudes of the transmitted

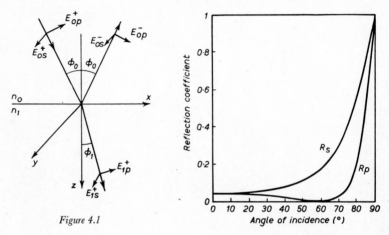

Figure 4.1

Figure 4.2 Variation of R_p, R_s with angle of incidence

waves to those of the incident waves. Our fears lest the law of the conservation of energy has failed are allayed when we consider, with the aid of Poynting's theorem, the energy in each medium. The energy is represented by the Poynting vector **S**.

$$\mathbf{S} = \frac{c}{4\pi} [\mathbf{E} \times \mathbf{H}] \qquad \text{.... 4(23)}$$

$$= \frac{c}{4\pi} n |\mathbf{E}|^2 \qquad \text{.... 4(24)}$$

where we consider propagation in a non-absorbing medium of refractive index n. (Equation 4(24) may be applied to propagation in an absorbing medium if n is replaced by $|n|$. See Section 4.3.)

The reflectances (defined as the ratios of reflected to transmitted energies) are simply given by

$$R_p = \frac{(E_{0p}^-)^2}{(E_{0p}^+)^2} = r_{1p}^2$$

and
$$R_s = \frac{(E_{0s}^-)^2}{(E_{0s}^+)^2} = r_{1s}^2$$

.... 4(25)

and the transmittances are given by

$$T_p = \frac{n_1(E_{1p}^+)^2}{n_0(E_{0p}^+)^2} = \frac{n_1}{n_0} t_{1p}^2$$

and
$$T_s = \frac{n_1(E_{1s}^+)^2}{n_0(E_{0s}^+)^2} = \frac{n_1}{n_0} t_{1s}^2$$

.... 4(26)

The variation of these coefficients with angle of incidence are shown in *Figure 4.2* for the case $n_0 = 1$, $n_1 = 1 \cdot 5$.

For normal incidence on an isotropic medium, the reflection and transmission coefficients, expressed in terms of refractive indices, become

$$R_p = R_s = \left(\frac{n_0 - n_1}{n_0 + n_1}\right)^2 \qquad \text{.... 4(27)}$$

$$T_p = T_s = \frac{4n_0 n_1}{(n_0 + n_1)^2} \qquad \text{.... 4(28)}$$

We may note in passing that with the use of Snell's law the Fresnel coefficients may be written as

$$r_{1p} = \frac{\tan(\varphi_1 - \varphi_0)}{\tan(\varphi_1 + \varphi_0)} \qquad \text{.... 4(29)}$$

$$t_{1p} = \frac{2 \sin \varphi_1 \cos \varphi_0}{\sin(\varphi_1 + \varphi_0) \cos(\varphi_1 - \varphi_0)} \qquad \text{.... 4(30)}$$

$$r_{1s} = \frac{\sin(\varphi_1 - \varphi_0)}{\sin(\varphi_1 + \varphi_0)} \qquad \text{.... 4(31)}$$

$$t_{1s} = \frac{2 \sin \varphi_1 \cos \varphi_0}{\sin(\varphi_1 + \varphi_0)} \qquad \text{.... 4(32)}$$

Although of use in dealing with problems involving reflection and transmission at a single surface, these forms are of little use in multilayer problems.

4.3. REFLECTION AT THE SURFACE OF AN ABSORBING MEDIUM

The equations of propagation of light in a transparent medium may, as indicated above, be taken over for the case of an absorbing medium by replacing the real refractive index by a complex term. The expressions for the Fresnel coefficients (equations 4(19)–4(22)) then also become complex and rather complicated. Replacing n_1 by $\boldsymbol{n}_1 = n_1 - ik_1$ we see that

$$\sin \varphi_1 = \frac{n_0 \sin \varphi_0}{n_1 - ik_1} \qquad \text{.... 4(33)}$$

so that φ_1 is complex and therefore does not represent the angle of refraction, except for the special case $\varphi_0 = \varphi_1 = 0$. For this case only, the Fresnel reflection coefficients (which are the same for both components of polarization) may be easily found.

$$r_{1p} = r_{1s} = \frac{n_0 - n_1 + ik_1}{n_0 + n_1 - ik_1} \qquad \text{.... 4(34)}$$

which gives, for the reflectance of the surface

$$\boldsymbol{R}_p = \boldsymbol{R}_s = \frac{(n_0 - n_1)^2 + k_1^2}{(n_0 + n_1)^2 + k_1^2} \qquad \text{.... 4(35)}$$

For other than normal incidence, exact expressions for the reflectance are cumbersome and approximations are used. References [2-4] give details of the types of approximation which are useful in this problem. For many absorbing materials, particularly metals in the visible region, $n^2 + k^2 \gg 1$. To this approximation the reflectances reduce to

$$\boldsymbol{R}_p = \frac{(n^2 + k^2) \cos^2 \varphi_0 - 2n \cos \varphi_0 + 1}{(n^2 + k^2) \cos^2 \varphi_0 + 2n \cos \varphi_0 + 1} \qquad \text{.... 4(36)}$$

$$\boldsymbol{R}_s = \frac{(n^2 + k^2) - 2n \cos \varphi_0 + \cos^2 \varphi_0}{(n^2 + k^2) + 2n \cos \varphi_0 + \cos^2 \varphi_0} \qquad \text{.... 4(37)}$$

The Fresnel transmission coefficients have no direct significance for the wave entering the absorbing medium since the attenuation of the wave depends on the distance travelled in the medium.

We may write the complex Fresnel reflection coefficients in the form

$$r_{1p} = \sigma_{1p} e^{i\beta_{1p}} \qquad \text{.... 4(38)}$$

$$r_{1s} = \sigma_{1s} e^{i\beta_{1s}} \qquad \text{.... 4(39)}$$

where σ_{1p}, σ_{1s} represent the (real) amplitudes of the reflected waves for unit incident amplitude and β_{1p}, β_{1s} are the phase changes at the surface. This form will be found of considerable use in dealing with the methods of determining optical constants by polarimetry (Section 5.7), since we may in these measurements conveniently determine the ratio $\sigma = \sigma_{1p}/\sigma_{1s}$ and the differential phase change

$\beta = \beta_{1p} - \beta_{1s}$. σ and β are related to the angle of incidence and the optical constants by the equation

$$\frac{1 + \sigma e^{i\beta}}{1 - \sigma e^{i\beta}} = \frac{\sin \varphi_0 \tan \varphi_0}{[(n - ik)^2 - \sin^2 \varphi_0]^{\frac{1}{2}}} \qquad \text{.... } 4(40)$$

(see e.g. reference[5]).

4.4. REFLECTION AND TRANSMISSION OF LIGHT BY A SINGLE FILM

(a) *Method of Summation*

We may apply the results of Section 4.3 to the determination of the reflection and transmission coefficients for a single non-absorbing layer, bounded on either side by semi-infinite non-absorbing layers. We do this by considering a beam incident on the film which is divided into reflected and transmitted parts. Such division occurs each time the beam strikes an interface so that the transmitted and reflected beams are obtained by summing the multiply-reflected and multiply-transmitted elements. For the case of the single layer and for this case only, the summation can be easily effected. The results are conveniently expressed in terms of Fresnel coefficients.

Suppose that a parallel beam of light of unit amplitude and of wavelength λ falls on a plane, parallel-sided, homogeneous, isotropic film of thickness d and refractive index n_1 supported on a substrate of index n_2 (*Figure 4.3*). The index of the first medium is n_0 and the angle of incidence in this medium φ_0.

We may write down the amplitudes of the successively reflected and transmitted beams in terms of the Fresnel coefficients, given by equations 4(19)–4(22). From the definitions of these coefficients, it is clear that the values of r and t for a given boundary depend on the direction of propagation of light across the boundary. Thus for normal incidence on a boundary between media of refractive indices n_0 and n_1 the Fresnel coefficient for reflection for light incident from n_0 is given by $(n_0 - n_1)/(n_0 + n_1)$ whilst that for the reverse direction is $(n_1 - n_0)/(n_1 + n_0)$. The corresponding Fresnel coefficients for transmission are $2n_0/(n_0 + n_1)$ for propagation from n_0 to n_1 and $2n_1/(n_0 + n_1)$ for propagation from n_1 to n_0. We may note in passing that certain workers (e.g. H. MAYER[6]) have taken no account of this difference, but have defined a single transmission coefficient, by expressions analogous to those given above and have used the same coefficient for propagation in both directions across the boundary. By compensating errors in applying the conservation of energy, the correct result has been obtained.

In treating the problem of the single layer, we shall denote the Fresnel coefficients for propagation from n_0 to n_1 by r_1 and t_1 as given by 4(19)–4(22). The corresponding coefficients for propagation from n_1 to n_0 will be written r_1' and t_1'. (Although to conform with the notation outlined at the beginning of the chapter, we should use + and − superscripts, there are reasons for not doing so. We

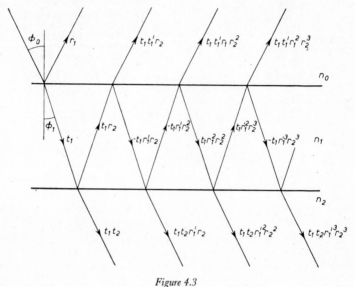

Figure 4.3

shall in the final expressions for the reflected and transmitted amplitudes always express the results in terms of r_1 and t_1; and, since the squares of these quantities are involved, the omission of the + superscript simplifies the symbol used.) The expressions given below will be valid for either direction of polarization provided that r and t are given the appropriate values from equations 4(19)–4(22). The second suffix (p or s) will therefore be omitted. From the form of the expression for the Fresnel reflection coefficient we see that r_1' is equal to $-r_1$.

The amplitudes of the successive beams reflected into medium n_0 are thus given by r_1, $t_1 t_1' r_2$, $-t_1 t_1' r_1 r_2^2$, $t_1 t_1' r_1^2 r_2^3$, ... and the transmitted amplitudes by $t_1 t_2$, $-t_1 t_2 r_1 r_2$, $t_1 t_2 r_1^2 r_2^2$, ... Writing δ_1, for the change in phase of the beam on traversing the film, we have

$$\delta_1 = \frac{2\pi}{\lambda} n_1 d_1 \cos \varphi_1 \qquad .... \; 4(41)$$

56

The reflected amplitude is thus given by

$$R = r_1 + t_1 t_1' r_2 e^{-2i\delta_1} - t_1 t_1' r_1 r_2^2 e^{-4i\delta_1} + \ldots .$$

$$= r_1 + \frac{t_1 t_1' r_2 e^{-2i\delta_1}}{1 + r_1 r_2 e^{-2i\delta_1}} \qquad \ldots . \ 4(42)$$

where the time-dependent factor is omitted. For non-absorbing media, this may be further simplified by writing the Fresnel transmission coefficients in terms of r_1, r_2. From conservation of energy (or from equations 4(19)–4(22)) we have

$$t_1 t_1' = 1 - r_1^2 \qquad \ldots . \ 4(43)$$

so that equation 4(42) becomes

$$R = \frac{r_1 + r_2 e^{-2i\delta_1}}{1 + r_1 r_2 e^{-2i\delta_1}} \qquad \ldots . \ 4(44)$$

The transmitted amplitude is given by

$$T = t_1 t_2 e^{-i\delta_1} - t_1 t_2 r_1 r_2 e^{-3i\delta_1} + t_1 t_2 r_1^2 r_2^2 e^{-5i\delta_1} - \ldots .$$

$$= \frac{t_1 t_2 e^{-i\delta_1}}{1 + r_1 r_2 e^{-2i\delta_1}} \qquad \ldots . \ 4(45)$$

Equations 4(44) and 4(45) are generally valid. For non-normal incidence, each takes two possible forms, depending on the state of polarization of the incident light. For light polarized with its electric vector parallel to the plane of incidence, the reflected and transmitted amplitudes are obtained by substituting for r_1, r_2, t_1 and t_2 from expressions corresponding to equations 4(19) and 4(20). For light polarized with the electric vector perpendicular to the plane of incidence, the Fresnel coefficients as given by equations 4(21) and 4(22) are used.

If the film is absorbing, or if it is bounded by absorbing media, then the values of n_0, n_1, n_2 are replaced by the corresponding complex quantities. The Fresnel coefficients then become complex and the values of R and T somewhat complicated, although still explicitly calculable. It must be remembered that these expressions give the amplitudes of the waves in the media bounding the film and that the *energies* of the corresponding beams are given by

$$n_0 R R^* = \frac{n_0(r_1^2 + 2r_1 r_2 \cos 2\delta_1 + r_2^2)}{(1 + 2r_1 r_2 \cos 2\delta_1 + r_1^2 r_2^2)} \qquad \ldots \ 4(46)$$

$$n_2 T T^* = \frac{n_2 t_1^2 t_2^2}{(1 + 2r_1 r_2 \cos 2\delta_1 + r_1^2 r_2^2)} \qquad \ldots \ 4(47)$$

Since we have considered a wave of unit *amplitude* (and not unit energy) in the first medium, the reflectance and transmittance (defined as ratios of reflected and transmitted energy to the incident energy) are given by

$$R = \frac{r_1^2 + 2r_1r_2 \cos 2\delta_1 + r_2^2}{1 + 2r_1r_2 \cos 2\delta_1 + r_1^2 r_2^2} \qquad \dots \ 4(48)$$

$$T = \frac{n_2}{n_0} \cdot \frac{t_1^2 t_2^2}{(1 + 2r_1r_2 \cos 2\delta_1 + r_1^2 r_2^2)} \qquad \dots \ 4(49)$$

The forms of the expressions for the amplitudes of the reflected and transmitted beams for the case of normal incidence are reasonably compact even when expressed in terms of the refractive indices. From equations 4(19)–4(22) we see that the Fresnel coefficients reduce to

$$r_1 = \frac{n_0 - n_1}{n_0 + n_1} \qquad\qquad t_1 = \frac{2n_0}{n_0 + n_1} \qquad \dots \ 4(50)$$

$$r_2 = \frac{n_1 - n_2}{n_1 + n_2} \qquad\qquad t_2 = \frac{2n_1}{n_1 + n_2} \qquad \dots \ 4(51)$$

so that equations 4(44) and 4(45) become

$$R = \frac{(n_0 - n_1)(n_1 + n_2)e^{i\delta_1} + (n_0 + n_1)(n_1 - n_2)e^{-i\delta_1}}{(n_0 + n_1)(n_1 + n_2)e^{i\delta_1} + (n_0 - n_1)(n_1 - n_2)e^{-i\delta_1}} \qquad \dots \ 4(52)$$

and

$$T = \frac{4n_0 n_1}{(n_0 + n_1)(n_1 + n_2)e^{i\delta_1} + (n_0 - n_1)(n_1 - n_2)e^{-i\delta_1}} \qquad \dots \ 4(53)$$

The reflectance and transmittance are given by

$$R = \frac{(n_0^2 + n_1^2)(n_1^2 + n_2^2) - 4n_0 n_1^2 n_2 + (n_0^2 - n_1^2)(n_1^2 - n_2^2) \cos 2\delta_1}{(n_0^2 + n_1^2)(n_1^2 + n_2^2) + 4n_0 n_1^2 n_2 + (n_0^2 - n_1^2)(n_1^2 - n_2^2) \cos 2\delta_1} \qquad \dots \ 4(54)$$

$$T = \frac{8n_0 n_1^2 n_2}{(n_0^2 + n_1^2)(n_1^2 + n_2^2) + 4n_0 n_1^2 n_2 + (n_0^2 - n_1^2)(n_1^2 - n_2^2) \cos 2\delta_1} \qquad \dots \ 4(55)$$

For non-absorbing media, these expressions are readily evaluated. If the film or the bounding media are absorbing, the values of n_0, n_1, n_2 need to be replaced by the complex $\boldsymbol{n} = n - ik$. The resulting expression is uselessly cumbersome to write out explicitly and is better evaluated by successive steps. Suitable programmes for calculating reflectance and transmittance are given in Section 4.9 below.

Returning to equations 4(44)–4(45), which give the reflected and transmitted amplitudes in terms of the Fresnel coefficients, we see that, for cases in which the Fresnel reflection coefficients are small enough for their products to be negligible compared with unity, the expressions reduce to

$$R \doteqdot r_1 + r_2 e^{-2i\delta_1} \qquad \text{.... } 4(56)$$

and

$$T \doteqdot (1 + r_1 + r_2)e^{-i\delta_1} \qquad \text{.... } 4(57)$$

and the corresponding expressions for reflectance and transmittance by

$$R = (r_1^2 + 2r_1 r_2 \cos 2\delta_1 + r_2^2) \qquad \text{.... } 4(58)$$

$$T = \frac{n_2}{n_0}(1 + r_1 + r_2)^2 \qquad \text{.... } 4(59)$$

(b) *Method using resultant waves*

It is clear that the method given above would become inordinately complicated if applied to the case of multiple layers. It has been applied to the system of two layers (A. W. CROOK[7]) and the form of the procedure for expressing the result for n layers in terms of that for $(n-1)$ layers has been determined. The type of reduction formulae involved are discussed in Section 4.6.

An alternative procedure to that of summing the individually-reflected and transmitted beams is to work with the vector sums of these waves and to apply the appropriate boundary conditions at the surfaces. Consider a wave incident on a film of index n_1 between media of indices n_0, n_2 as before (*Figure 4.4*) and suppose

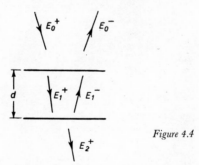

Figure 4.4

the amplitude of the electric vector is E_0^+. That of the reflected wave is E_0^-. Inside the film the resultants of all the positive-going waves sum to E_1^+ and those of the negative-going to E_1^-. In the third medium we shall have only the positive-going, transmitted wave of amplitude E_2^+. For non-normal incidence on an isotropic film, where it is necessary to distinguish the plane of polarization, we do this by adding the suffix p or s in accordance with the discussion in (a) above. As before, we choose the z-axis normal to the plane of the film and the plane of incidence as the xz plane. We take the origin of co-ordinates as the point of incidence in the n_0/n_1 surface.

The phase factors associated with the waves in the three media are of the form

$$\exp i\left(\omega t - \frac{2\pi n_0 x \sin \varphi_0}{\lambda} \pm \frac{2\pi n_0 z \cos \varphi_0}{\lambda}\right) \quad \text{(medium } n_0\text{)}$$

$$\exp i\left(\omega t - \frac{2\pi n_1 x \sin \varphi_1}{\lambda} \pm \frac{2\pi n_1 z \cos \varphi_1}{\lambda}\right) \quad \text{(medium } n_1\text{)}$$

$$\exp i\left(\omega t - \frac{2\pi n_2 x \sin \varphi_2}{\lambda} - \frac{2\pi n_2 z \cos \varphi_2}{\lambda}\right) \quad \text{(medium } n_2\text{)}$$

where the upper sign indicates the wave travelling in the negative direction of z. We write down the z-dependent parts of the x- and y- components of the electric and magnetic vectors and apply the boundary conditions at each of the interfaces. Denoting $(2\pi n_m \cos \varphi_m)/\lambda$ by $\varkappa_m$ we have

$$\left.\begin{aligned}
E_{0x} &= (E_{0p}^+ e^{-i\varkappa_0 z} + E_{0p}^- e^{+i\varkappa_0 z}) \cos \varphi_0 \\
E_{0y} &= E_{0s}^+ e^{-i\varkappa_0 z} + E_{0s}^- e^{+i\varkappa_0 z} \\
H_{0x} &= (-E_{0s}^+ e^{-i\varkappa_0 z} + E_{0s}^- e^{+i\varkappa_0 z}) n_0 \cos \varphi_0 \\
H_{0y} &= (E_{0p}^+ e^{-i\varkappa_0 z} - E_{0p}^- e^{+i\varkappa_0 z}) n_0
\end{aligned}\right\} \quad \dots\ 4(60)$$

$$\left.\begin{aligned}
E_{1x} &= (E_{1p}^+ e^{-i\varkappa_1 z} + E_{1p}^- e^{+i\varkappa_1 z}) \cos \varphi_1 \\
E_{1y} &= E_{1s}^+ e^{-i\varkappa_1 z} + E_{1s}^- e^{+i\varkappa_1 z} \\
H_{1x} &= (-E_{1s}^+ e^{-i\varkappa_1 z} + E_{1s}^- e^{+i\varkappa_1 z}) n_1 \cos \varphi_1 \\
H_{1y} &= (E_{1p}^+ e^{-i\varkappa_1 z} - E_{1p}^- e^{+i\varkappa_1 z}) n_1
\end{aligned}\right\} \quad \dots\ 4(61)$$

For the third medium, there is no wave in the negative direction so that

$$\left.\begin{aligned}
E_{2x} &= E_{2p}^+ e^{-i\varkappa_2 z} \cos \varphi_2 \\
E_{2y} &= E_{2s}^+ e^{-i\varkappa_2 z} \\
H_{2x} &= -E_{2s}^+ e^{-i\varkappa_2 z} n_2 \cos \varphi_2 \\
H_{2y} &= E_{2p}^+ e^{-i\varkappa_2 z} n_2
\end{aligned}\right\} \quad \dots\ 4(62)$$

Equating the tangential components of the electric and magnetic vectors for the planes $z = 0$ and $z = d_1$ we obtain

$$\left.\begin{aligned}
(E_{0p}^+ + E_{0p}^-) \cos \varphi_0 &= (E_{1p}^+ + E_{1p}^-) \cos \varphi_1 \\
(E_{0p}^+ - E_{0p}^-) n_0 &= (E_{1p}^+ - E_{1p}^-) n_1
\end{aligned}\right\} \quad \dots\ 4(63)$$

$$\left.\begin{aligned} E_{0s}^+ + E_{0s}^- &= E_{1s}^+ + E_{1s}^- \\ (-E_{0s}^+ + E_{0s}^-)n_0 \cos\varphi_0 &= (-E_{1s}^+ + E_{1s}^-)n_1 \cos\varphi_1 \end{aligned}\right\} \quad \dots 4(64)$$

$$\left.\begin{aligned} (E_{1p}^+ e^{-i\varkappa_1 d_1} + E_{1p}^- e^{+i\varkappa_1 d_1})\cos\varphi_1 &= E_{2p}^+ e^{-i\varkappa_2 d_1}\cos\varphi_2 \\ (E_{1p}^+ e^{-i\varkappa_1 d_1} - E_{1p}^- e^{+i\varkappa_1 d_1})n_1 &= E_{2p}^+ e^{-i\varkappa_2 d_1}\, n_2 \end{aligned}\right\} \dots 4(65)$$

$$\left.\begin{aligned} E_{1s}^+ e^{-i\varkappa_1 d_1} + E_{1s}^- e^{+i\varkappa_1 d_1} &= E_{2s}^+ e^{-i\varkappa_2 d_1} \\ (-E_{1s}^+ e^{-i\varkappa_1 d_1} + E_{1s}^- e^{+i\varkappa_1 d_1})\,n_1\cos\varphi_1 &= E_{2s}^+ e^{-i\varkappa_2 d_1}\, n_2 \cos\varphi_2 \end{aligned}\right\} \dots 4(66)$$

It is thus seen that linear relations connect the components of the electric vector in one medium with those of the corresponding vectors in the preceding medium—a fact which enables the solution of the problem of multiple layers to be dealt with simply and elegantly (see Section 4.8). The relations between vectors in successive layers are conveniently expressed in terms of Fresnel coefficients, following the treatment of F. Abelès[1]. It is found that the form of the relation between the vectors polarized in the plane of incidence is exactly the same as that for the vectors polarized in the perpendicular plane. The suffixes p and s may therefore be dropped although it must be remembered that the values of the Fresnel coefficients depend on the state of polarization. In applying the equations given below, the Fresnel coefficient appropriate to the plane of polarization under consideration must be used (equations $4(19)$–$4(22)$). We then have

$$\left.\begin{aligned} E_0^+ &= \frac{1}{t_1}\, E_1^+ + \frac{r_1}{t_1}\, E_1^- \\ E_0^- &= \frac{r_1}{t_1}\, E_1^+ + \frac{1}{t_1}\, E_1^- \end{aligned}\right\} \quad \dots 4(67)$$

and

$$\left.\begin{aligned} E_1^+ e^{-i\varkappa_1 d_1} &= \frac{1}{t_2}\, E_2^+ e^{-i\varkappa_2 d_1} \\ E_1^- e^{-i\varkappa_1 d_1} &= \frac{r_2}{t_2}\, E_2^- e^{-i\varkappa_2 d_1} \end{aligned}\right\} \quad \dots 4(68)$$

These relations allow the amplitudes of the light in each medium and in each direction to be determined in terms of E_0^+ and hence the reflection and transmission coefficients to be found. Writing δ_1 for the phase change on traversing the film once, i.e.

$$\delta_1 = \frac{2\pi}{\lambda}\, n_1 d_1 \cos\varphi_1 \quad \text{(see equation } 4(41)\text{)}$$

we obtain
$$E_0^- = \frac{r_1 + r_2 e^{-2i\delta_1}}{1 + r_1 r_2 e^{-2i\delta_1}}\, E_0^+ \qquad \dots 4(69)$$

and $\quad\quad E_2^+ e^{-i\varkappa_2 d_1} = \dfrac{t_1 t_2 e^{-i\delta_1}}{1 + r_1 r_2 e^{-2i\delta_1}}\, E_0^+$ $\quad$ .... 4(70)

in agreement with equations 4(44)–4(45). The reflected and transmitted energies may then be determined as before and lead to equations 4(46) and 4(47). (The phase factor multiplying the amplitude E_2^+ results from the choice of origin. It disappears when the transmitted energy is calculated.)

Although no shorter than the method of writing down the amplitudes of the multiply-reflected beams, this treatment is very readily adapted to the problem of multiple layers.

4.5. Reflection and Transmission of Light by Two Films

In view of the growth in importance of double-layer films (see e.g. Chapter 7), we shall give the results for the reflection and transmission coefficients for such a system. The system is as shown in

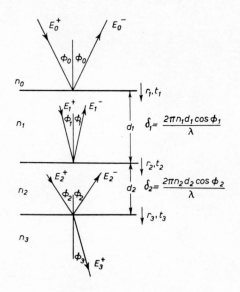

Figure 4.5   Two films

Figure 4.5. There are several ways in which the result for two films may be obtained. We may repeat the process outlined in Section 4.4 above, specifying the light vectors in the various media and applying boundary conditions at each of the interfaces. This constitutes in effect an application of the general method (Section 4.8), the general results for this method being obtained by precisely this process. Or we may follow the method given by P. ROUARD[8]. Treating the problem in two stages, the Fresnel coefficients are first obtained for the film (n_2) +substrate system using expressions of the form 4(69) and 4(70). Thus the Fresnel reflection coefficient for the system is given by $(r_2 + r_3 e^{-2i\delta_2})/(1 + r_2 r_3 e^{-2i\delta_2})$ and the transmission coefficient by $(t_2 t_3 e^{-i\delta_2})/(1 + r_2 r_3 e^{-2i\delta_2})$, where the r's and t's are the Fresnel coefficients for the two surfaces, as defined by equations 4(19)–4(22). We now consider the n_1 layer as covering a surface for which these are the Fresnel coefficients. The reflection and transmission coefficients are given by equations 4(69) and 4(70) where we must replace r_2 and t_2 by the coefficients given above.

Yet a third possibility is to consider the multiply-reflected beams, as for the one-layer case. This has been done by A. W. Crook[7].

The results obtained are as follows, where $\delta_r = \dfrac{2\pi}{\lambda} n_r d_r \cos \varphi_r$

$$E_0^- = \frac{r_1 + r_2 e^{-2i\delta_1} + r_3 e^{-2i(\delta_1 + \delta_2)} + r_1 r_2 r_3 e^{-2i\delta_2}}{1 + r_1 r_2 e^{-2i\delta_1} + r_1 r_3 e^{-2i(\delta_1 + \delta_2)} + r_2 r_3 e^{-2i\delta_2}} E_0^+ \quad \dots\ 4(71)$$

$$E_3^+ e^{-i\varkappa_3 d_2} = \frac{t_1 t_2 t_3 e^{-i(\delta_1 + \delta_2)}}{1 + r_1 r_2 e^{-2i\delta_1} + r_1 r_3 e^{-2i(\delta_1 + \delta_2)} + r_2 r_3 e^{-2i\delta_2}} E_0^+ \quad \dots\ 4(72)$$

4.6. THE EXTENSION TO A SYSTEM OF MULTIPLE LAYERS

(a) *Rouard's treatment.*   *Transparent layers*

Methods by which the derivation of the reflection coefficient for a single layer may be extended to the case of any number of layers have been given by P. Rouard[8], A. W. Crook[7] and A. VAŠÍČEK[9]. There are two possible approaches. Since a single film bounded by two surfaces possesses an effective reflection coefficient and accompanying phase change, then such a film may be replaced by a single surface with these properties. Rouard starts with the film next to the supporting substrate and works step by step through the intervening layers to the top of the system. Vašíček starts with the top layer and moves downwards towards the substrate. The expression for the reflectivity of the system is slightly more tractable when Rouard's scheme as in *Figure 4.6* is used. Starting with two films with Fresnel coefficients r_1, r_2, r_3 (for light travelling from n_0 to n_2), we first compute the amplitude and phase of the light reflected by the lower film (d_2). Writing ϱ_2 for the real amplitude and Δ_2 for the phase, we have

$$\varrho_2 e^{i\Delta_2} = \frac{r_2 + r_3 e^{-2i\delta_2}}{1 + r_2 r_3 e^{-2i\delta_2}} \qquad \ldots 4(73)$$

This is the effective Fresnel coefficient for the layer n_2 and is inserted into the corresponding expression for the Fresnel coefficient for the whole system which is now regarded as a film of thickness d_1 lying on a surface whose Fresnel coefficient is $\varrho_2 e^{i\Delta_2}$. We thus have

$$\varrho_1 e^{i\Delta_1} = \frac{r_1 + \varrho_2 e^{i\Delta_2} e^{-2i\delta_1}}{1 + r_1 \varrho_2 e^{i\Delta_2} e^{-2i\delta_1}} \qquad \ldots 4(74)$$

and on eliminating ϱ_2, Δ_2 between 4(73) and 4(74), we obtain

$$\varrho_1 e^{i\Delta_1} = \frac{r_1 + r_2 e^{-2i\delta_1} + r_3 e^{-2i(\delta_1 + \delta_2)} + r_1 r_2 r_3 e^{-2i\delta_1}}{1 + r_1 r_2 e^{-2i\delta_1} + r_1 r_3 e^{-2i(\delta_1 + \delta_2)} + r_2 r_3 e^{-2i\delta_2}} \qquad \ldots 4(75)$$

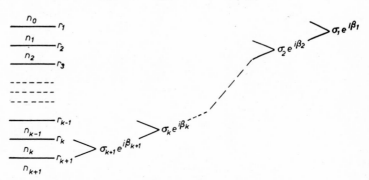

Figure 4.6   Rouard's scheme for dealing with multiple films

Consider now a system of k layers (*Figure 4.6*). The lowest layer (n_k) has an effective Fresnel coefficient given by

$$\varrho_k e^{i\Delta_k} = \frac{r_k + r_{k+1} e^{-2i\delta_k}}{1 + r_k r_{k+1} e^{-2i\delta_k}} \qquad \ldots 4(76)$$

On forming the product of this expression with its complex conjugate, we obtain

$$\varrho_k^2 = \frac{r_k^2 + r_{k+1}^2 + 2r_k r_{k+1} \cos 2\delta_k}{1 + r_k^2 r_{k+1}^2 + 2r_k r_{k+1} \cos 2\delta_k} \qquad \ldots 4(77)$$

Writing Δ_k in the form

$$\Delta_k = \eta_k - \xi_k \qquad \ldots 4(78)$$

we see from 4(76) that

$$\left.\begin{aligned}
\tan \xi_k &= \frac{r_{k+1} \sin 2\delta_k}{r_k + r_{k+1} \cos 2\delta_k} \\
\tan \eta_k &= \frac{r_k r_{k+1} \sin 2\delta_k}{1 + r_k r_{k+1} \cos 2\delta_k}
\end{aligned}\right\} \qquad \dots 4(79)$$

so that ξ_k, η_k and therefore Δ_k may be easily found from the Fresnel coefficients at each interface and from the optical thickness of the film.

The $(k-1)^{th}$ layer is then added giving an effective Fresnel coefficient

$$\varrho_{k-1} e^{i\Delta_{k-1}} = \frac{r_{k-1} + \varrho_k e^{i\Delta_k} e^{-2i\delta_{k-1}}}{1 + r_{k-1} \varrho_k e^{i\Delta_k} e^{-2i\delta_{k-1}}} \qquad \dots 4(80)$$

ϱ_{k-1} and Δ_{k-1} are calculated in the same way as for ϱ_k and Δ_k and the process is repeated successively until the final coefficient ϱ_1 is obtained, for the whole system. The reflection coefficient of the system is then simply given by ϱ_1^2 (since ϱ_1 is the *real* amplitude).

(b) *Absorbing layers*

The forms of the expressions given above are useful only in the case of non-absorbing layers, for which the Fresnel coefficients are real. A system of absorbing layers, for which the refractive indices are replaced by complex quantities and for which, therefore, the Fresnel coefficients are complex, may be dealt with by the following method. The general case, for non-normal incidence, is excessively cumbersome and is more suitably dealt with by the method of Abelès[1]. For *normal incidence*, we proceed as follows.

If the kth and $(k-1)^{th}$ layers are absorbing, with complex indices $\boldsymbol{n}_k \equiv n_k - ik_k$ and $\boldsymbol{n}_{k-1} \equiv n_{k-1} - ik_{k-1}$ then the Fresnel coefficient for light travelling from $(k-1)$ to k is given by

$$r_k = g_k + ih_k \equiv \frac{\boldsymbol{n}_{k-1} - \boldsymbol{n}_k}{\boldsymbol{n}_{k-1} + \boldsymbol{n}_k} \qquad \dots 4(81)$$

whence

$$g_k = \frac{n_{k-1}^2 - n_k^2 + k_{k-1}^2 - k_k^2}{(n_{k-1} + n_k)^2 + (k_{k-1} + k_k)^2} \qquad \dots 4(82a)$$

$$h_k = \frac{2(n_{k-1} k_k - n_k k_{k-1})}{(n_{k-1} + n_k)^2 + (k_{k-1} + k_k)^2} \qquad \dots 4(82b)$$

These values are inserted in 4(76) together with the (complex) value of δ_k; the exponential is written

$$\exp(-2i\delta_k) = \exp\left(-\frac{4\pi}{\lambda}k_k d_k\right)\exp\left(-\frac{4\pi}{\lambda}in_k d_k\right)$$

$$= e^{-a_k}(\cos\theta_k - i\sin\theta_k)$$

We then obtain

$$\varrho_k^2 = \frac{\{g_k + e^{-a_k}(g_{k+1}\cos\theta_k + h_{k+1}\sin\theta_k)\}^2 + \{h_k + e^{-a_k}(h_{k+1}\cos\theta_k - g_{k+1}\sin\theta_k)\}^2}{\{1 + e^{-a_k}[(g_k g_{k+1} - h_k h_{k+1})\cos\theta_k + (h_k g_{k+1} + h_{k+1}g_k)\sin\theta_k]\}^2 + \{e^{-a_k}[(h_k g_{k+1} + h_{k+1}g_k)\cos\theta_k - (g_k g_{k+1} - h_k h_{k+1})\sin\theta_k]\}^2}$$

.... 4(83)

and, writing $\Delta_k = \eta_k - \xi_k$ as before, we obtain

$$\tan\xi_k = \frac{e^{-a_k}[(h_k g_{k+1} + h_{k+1}g_k)\cos\theta_k - (g_k g_{k+1} - h_k h_{k+1})\sin\theta_k]}{1 + e^{-a_k}[(g_k g_{k+1} - h_k h_{k+1})\cos\theta_k - (h_k g_{k+1} + h_{k+1}g_k)\sin\theta_k]}$$

and

$$\left.\tan\eta_k = \frac{h_k + e^{-a_k}(h_{k+1}\cos\theta_k - g_{k+1}\sin\theta_k)}{g_k + e^{-a_k}(g_{k+1}\cos\theta_k + h_{k+1}\sin\theta_k)}\right\} \qquad \text{.... 4(84)}$$

These values of ϱ_k, Δ_k are then inserted in equation 4(80) and the process repeated successively as before.

The form of the equations resulting from this method of analysis is such that the dependence of the properties of a series of layers of given materials on the variation in thicknesses is reasonably straight-forward. For layers with given optical constants the values of g_k, h_k (equation 4(82)) are fixed. It is seen from equations 4(83) and 4(84) that many of the terms involve products of g's and h's only so that, in investigating the variation of properties with film thick-nesses, these quantities would remain unchanged. Change in the thicknesses of the layers merely changes the values of a_k and θ_k. In certain cases—e.g. very thin metal films—the crystalline structure, and in consequence the optical constants, show marked variation with film thickness. This must be taken into account in computing the properties of multiple layer systems. The variation in optical constants may be quite violent so that computations made using the values characteristic of the bulk material would be wildly awry.

4.7. Optical Impedance

By introducing the concept of optical impedance and admittance, equations may be obtained for the reflectance of a system of layers which are analogous to the equations appropriate to the electrical properties of transmission lines. The optical impedance must be defined so that it varies continuously with the thickness of the system of layers, even through discontinuities in refractive index. The boundary conditions at the interfaces of the system provide us with a suitable variable. Since the tangential components of the electric and magnetic vectors are continuous at the boundaries, we may define the optical impedance Z as

$$Z = \frac{E_{\text{tangential}}}{H_{\text{tangential}}} = \frac{1}{Y} \qquad \qquad \dots \ 4(85)$$

where Y is the optical admittance. Y is found to satisfy a differential equation which may be easily solved under certain conditions.

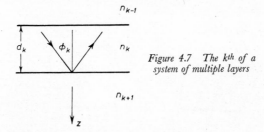

Figure 4.7   The kth of a
system of multiple layers

The principle of this method is to express the tangential components of the electric and magnetic vectors on one side of a film in terms of those on the opposite side. Since the tangential components are continuous at the boundaries, this process may be repeated through a whole system of films. A linear relation is found to exist between the tangential components of E and H on opposite sides of a film.

Consider the kth film of a system (*Figure 4.7*). Then, with the notation of Section 4.4 and the same co-ordinate system, we may write down the tangential components of E and H on each side of the film. We shall consider the two possible cases of polarization separately.

(i) *Electric vector parallel to plane of incidence*

The tangential components are then E_{kx} and H_{ky} where

$$\left. \begin{array}{l} E_{kx} = (E_{kp}^{+} e^{-i\varkappa_k z} + E_{kp}^{-} e^{+i\varkappa_k z}) \cos \varphi_k \\ H_{ky} = (E_{kp}^{+} e^{-i\varkappa_k z} - E_{kp}^{-} e^{+i\varkappa_k z}) n_k \end{array} \right\} \qquad \dots \ 4(86)$$

We may choose the plane $z = 0$ on the left-hand side of the kth layer and we then write down the values of E_{kx}, H_{xy} on this plane

67

and on the plane $z = d_k$. On eliminating E_{kp}^+ and E_{kp}^- from the resulting equations, we obtain the following relations between the tangential components. To simplify the notation, we write E_{k-1} for the tangential component of E on the plane separating $(k-1)$ from (k) (i.e. $z = 0$) and E_k for the plane $z = d_k$

$$\left.\begin{aligned}
E_{k-1} &= E_k \cos \varkappa_k d_k + \frac{i \cos \varphi_k}{n_k} H_k \sin \varkappa_k d_k \\
H_{k-1} &= \frac{i n_k}{\cos \varphi_k} E_k \sin \varkappa_k d_k + H_k \cos \varkappa_k d_k
\end{aligned}\right\} \quad \dots 4(87)$$

(ii) *Electric vector perpendicular to plane of incidence*

For this case we have

$$\left.\begin{aligned}
E_{ky} &= (E_{ks}^+ e^{-i\varkappa_k z} + E_{ks}^- e^{+i\varkappa_k z}) \\
H_{kx} &= (-E_{ks}^+ e^{-i\varkappa_k z} + E_{ks}^- e^{+i\varkappa_k z}) n_k \cos \varphi_k
\end{aligned}\right\} \quad \dots 4(88)$$

Following the procedure given above for the other plane of polarization, we obtain

$$\left.\begin{aligned}
E_{k-1} &= E_k \cos \varkappa_k d_k - \frac{i}{n_k \cos \varphi_k} H_k \sin \varkappa_k d_k \\
H_{k-1} &= -i n_k \cos \varphi_k E_k \sin \varkappa_k d_k + H_k \cos \varkappa_k d_k
\end{aligned}\right\} \quad \dots 4(89)$$

We may express 4(87) and 4(89) in a single form by introducing the characteristic admittance (η) of the medium, defined as the ratio of the tangential components of the magnetic to electric vector for the positive-going wave. For case (i) above, we have $\eta_{k\parallel} = n_k / \cos \varphi_k$ while for case (ii) $\eta_{k\perp} = -n_k \cos \varphi_k$ (from equations 4(86) and 4(88), ignoring the E_{kp}^- and E_{ks}^- terms). We may then write

$$E_{k-1} = E_k \cos \varkappa_k d_k + \frac{i \sin \varkappa_k d_k}{\eta_k} H_k \qquad \dots 4(90a)$$

$$H_{k-1} = i \eta_k E_k \sin \varkappa_k d_k + H_k \cos \varkappa_k d_k \qquad \dots 4(90b)$$

where the value of η_k appropriate to the plane of polarization under consideration is inserted. The reflectance at any interface in the system is readily written down in terms of the characteristic admittance of the medium η_k and the admittance at any point Y_k. From equation 4(86), putting $R_k = E_{kp}^- / E_{kp}^+$ we see that

$$R_k = \frac{\eta_k - Y_k}{\eta_k + Y_k} \qquad \dots 4(91)$$

4(91) gives the (complex) amplitude reflection coefficient. The reflectance is given by $R_k R_k^*$.

Determination of η, Υ thus enable the reflectance to be calculated. The value of Υ is given by dividing 4(90b) by 4(90a) whence

$$\Upsilon_{k-1} = \eta_k \cdot \frac{\Upsilon_k \cos \varkappa_k d_k + i\eta_k \sin \varkappa_k d_k}{\eta_k \cos \varkappa_k d_k + i\Upsilon_k \sin \varkappa_k d_k} \qquad \dots 4(92)$$

Thus for a system of homogeneous layers, the admittances at each boundary are easily found.

The above treatment may be readily extended to deal with the problem of layers in which the refractive index varies in the direction of propagation. Suppose that, at a point in a system of layers at which the admittance is Υ, we add a thin layer, dl, of characteristic admittance η. The value of Υ changes to $\Upsilon + d\Upsilon$ and, from 4(92) we see that, if $dq = \varkappa dl$, then

$$\frac{d\Upsilon}{dq} = i\eta\left(\frac{\Upsilon^2}{\eta^2} - 1\right) \qquad \dots 4(93a)$$

Or, putting $\zeta = \Upsilon/\eta$

$$\frac{d\zeta}{dq} = i(\zeta^2 - 1) - \zeta\frac{d}{dq}(ln\eta) \qquad \dots 4(93b)$$

For a system, then, in which the refractive index varies in a known manner with thickness, so that $(d/dq)(ln\eta)$ is known, the solution of equation 4(93b) yields Υ from which the reflectance of the system may be found.

B. S. BLAISSE[10] has applied this method to the case of a layer in which the reciprocal of the refractive index is a linear function of distance across the film, for which case the solution is simple, since $(d/dq)(ln\eta)$ is constant. He has also applied the method, with the help of a graphical construction, to the case of a film in which the refractive index varies linearly with distance. Such films have been made by evaporation (J. STRONG[11]) and are of importance in that their reflection-reducing properties surpass those of corresponding systems of homogeneous films (see Chapter 7).

4.8. MATRIX METHOD USING FRESNEL COEFFICIENTS

We may notice that equations 4(90) representing the linear dependence of E_{k-1}, H_{k-1} on E_k, H_k, may be written in matrix form

$$\begin{pmatrix} E_{k-1} \\ H_{k-1} \end{pmatrix} = \begin{pmatrix} \cos \varkappa_k d_k & i\dfrac{\sin \varkappa_k d_k}{\eta_k} \\ i\eta_k \sin \varkappa_k d_k & \cos \varkappa_k d_k \end{pmatrix} \begin{pmatrix} E_k \\ H_k \end{pmatrix} \qquad \dots 4(94)$$

The components E_0, H_0 may thus be expressed in terms of E_k, H_k and hence the transmission coefficient of the system determined.

A form slightly more convenient for computation is obtained if the relations between the electric vectors in successive layers are expressed in terms of Fresnel coefficients, as is done by Abelès[1].

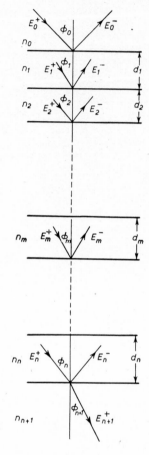

Figure 4.8

With the notation of Section 4.4 applied to the system of n layers shown in *Figure 4.8*, we obtain for the x and y components of E and H in the mth layer

$$
\left.
\begin{aligned}
E_{mx} &= (E_{mp}^{+}e^{-i\varkappa_m z} + E_{mp}^{-}e^{+i\varkappa_m z})\cos\varphi_m \\
E_{my} &= E_{ms}^{+}e^{-i\varkappa_m z} + E_{ms}^{-}e^{+i\varkappa_m z} \\
H_{mx} &= (-E_{ms}^{+}e^{-i\varkappa_m z} + E_{ms}^{-}e^{+i\varkappa_m z})n_m\cos\varphi_m \\
H_{my} &= (E_{mp}^{+}e^{-i\varkappa_m z} - E_{mp}^{-}e^{+i\varkappa_m z})n_m
\end{aligned}
\right\}
\quad \ldots 4(95)
$$

70

where

$$\varkappa_m = \frac{2\pi n_m \cos \varphi_k}{\lambda}$$

Writing $c_m = \sum\limits_{i=1}^{m-1} d_i$ we obtain, for the mth surface, separating the

$(m-1)^{th}$ and the mth layers.

$$(E^+_{m-1,p}e^{-i\varkappa_{m-1}c_m} + E^-_{m-1,p}e^{+i\varkappa_{m-1}c_m}) \cos \varphi_{m-1}$$
$$= (E^+_{mp}e^{-i\varkappa_m c_m} + E^-_{mp}e^{+i\varkappa_m c_m}) \cos \varphi_m \quad .... \quad 4(96a)$$

$$n_{m-1}(E^+_{m-1,p}e^{-i\varkappa_{m-1}c_m} - E^-_{m-1,p}e^{+i\varkappa_{m-1}c_m})$$
$$= n_m(E^+_{mp}e^{-i\varkappa_m c_m} - E^-_{mp}e^{+i\varkappa_m c_m}) \quad .... \quad 4(96b)$$

for the p component and

$$E^+_{m-1,s}e^{-i\varkappa_{m-1}c_m} + E^-_{m-1,s}e^{+i\varkappa_{m-1}c_m} = E^+_{ms}e^{-i\varkappa_m c_m} + E^-_{ms}e^{+i\varkappa_m c_m} \quad .... \quad 4(97a)$$

$$(-E^+_{m-1,s}e^{-i\varkappa_m c_m} + E^-_{m-1,s}e^{+i\varkappa_m c_m})n_{m-1} \cos \varphi_{m-1}$$
$$= (-E^+_{ms}e^{-i\varkappa_m c_m} + E^-_{ms}e^{+i\varkappa_m c_m})n_m \cos \varphi_m \quad ... \quad 4(97b)$$

for the s component.

These equations may be written in terms of the Fresnel coefficients. The forms of the resulting equations for the p and s components are identical so that the suffix p or s from the Fresnel coefficients may be dropped. It must, however, be remembered that the values of the Fresnel coefficients depend on the plane of polarization considered and that the appropriate values, from equations 4(19) to 4(22) must be inserted.

We obtain, from equations 4(96) and 4(97) using the values of r, t as defined in Section 4.2

$$E^+_{m-1}e^{-i\varkappa_{m-1}c_m} = (E^+_m e^{-i\varkappa_m c_m} + r_m E^-_m e^{+i\varkappa_m c_m})/t_m \quad .... \quad 4(98a)$$
$$E^-_{m-1}e^{+i\varkappa_{m-1}c_m} = (r_m E^+_m e^{-i\varkappa_m c_m} + E^-_m e^{+i\varkappa_m c_m})/t_m \quad .... \quad 4(98b)$$

By writing $\delta_m = \varkappa_m c_m$ and a change of origin, equations 4(98) may be further simplified to

$$E^+_{m-1} = (e^{i\delta_{m-1}}E^+_m + r_m e^{i\delta_{m-1}}E^-_m)/t_m \quad .... \quad 4(99a)$$
$$E^-_{m-1} = (r_m e^{-i\delta_{m-1}}E^+_m + e^{-i\delta_{m-1}}E^-_m)/t_m \quad .... \quad 4(99b)$$

which recurrence relation may be written in matrix form

$$\begin{pmatrix} E^+_{m-1} \\ E^-_{m-1} \end{pmatrix} = \frac{1}{t_m} \begin{pmatrix} e^{i\delta_{m-1}} & r_m e^{i\delta_{m-1}} \\ r_m e^{-i\delta_{m-1}} & e^{-i\delta_{m-1}} \end{pmatrix} \begin{pmatrix} E^+_m \\ E^-_m \end{pmatrix} \quad .... \quad 4(100)$$

For a system of n layers (*Figure 4.8*) we require to know the relation between E^+_{n+1} and E^+_0 so that the transmission coefficient may be obtained; and also that between E_0^- and E_0^+ for the reflection coefficient. From 4(100) we obtain

$$\begin{pmatrix} E_0^+ \\ E_0^- \end{pmatrix} = \frac{(C_1)(C_2) \ \ldots\ldots\ (C_{n+1})}{t_1 t_2 \ \ldots\ldots\ t_{n+1}} \begin{pmatrix} E_{n+1}^+ \\ E_{n+1}^- \end{pmatrix} \qquad \ldots\ 4(101)$$

where

$$(C_m) = \begin{pmatrix} e^{i\delta_{m-1}} & r_m e^{i\delta_{m-1}} \\ r_m e^{-i\delta_{m-1}} & e^{-i\delta_{m-1}} \end{pmatrix} \qquad \ldots\ 4(102)$$

We may therefore express E_0^- and E_{n+1}^+ in terms of E_0^+ and so obtain the reflection and transmission coefficients. We note, however, that since there is no negative-going wave in the $(n+1)^{th}$ medium, we put $E_{n+1}^- = 0$. Writing the matrix product

$$(C_1)(C_2) \ldots\ldots (C_{n+1}) = \begin{pmatrix} a & b \\ c & d \end{pmatrix} \qquad \ldots\ 4(103)$$

we obtain, from 4(101)

$$R = \frac{E_0^-}{E_0^+} = \frac{c}{a} \quad \text{and} \quad T = \frac{E_{n+1}^+}{E_0^+} = \frac{t_1 t_2 \ldots t_{n+1}}{a} \qquad \ldots\ 4(104)$$

The reflectance and transmittance are given by

$$\boldsymbol{R} = \frac{(E_0^-)(E_0^-)^*}{(E_0^+)(E_0^+)^*} \quad \text{and} \quad \boldsymbol{T} = \frac{n_{n+1}}{n_0} \frac{(E_{n+1}^+)(E_{n+1}^+)^*}{(E_0^+)(E_0^+)^*} \qquad \ldots\ 4(105)$$

so that we obtain

$$\boldsymbol{R} = \frac{cc^*}{aa^*} \qquad \ldots\ 4(106a)$$

$$\boldsymbol{T} = \frac{(t_1 t_2 \ldots t_{n+1})(t_1^* t_2^* \ldots t_{n+1}^*)}{aa^*} \qquad \ldots\ 4(106b)$$

For a single film we have

$$\begin{pmatrix} a & b \\ c & d \end{pmatrix} = \begin{pmatrix} 1 & r_1 \\ r_1 & 1 \end{pmatrix} \begin{pmatrix} e^{i\delta_1} & r_2 e^{i\delta_1} \\ r_2 e^{-i\delta_1} & e^{-i\delta_1} \end{pmatrix}$$

$$= \begin{pmatrix} e^{i\delta_1} + r_1 r_2 e^{-i\delta_1} & r_2 e^{i\delta_1} + r_1 e^{-i\delta_1} \\ r_1 e^{i\delta_1} + r_2 e^{-i\delta_1} & r_1 r_2 e^{i\delta_1} + e^{-i\delta_1} \end{pmatrix} \qquad \ldots\ 4(107)$$

So that, from equation 4(104) we have

$$R = \frac{r_1 e^{i\delta_1} + r_2 e^{-i\delta_1}}{e^{i\delta_1} + r_1 r_2 e^{-i\delta_1}} \qquad \ldots\ 4(108a)$$

$$T = \frac{t_1 t_2}{e^{i\delta_1} + r_1 r_2 e^{-i\delta_1}} \qquad \ldots\ 4(108b)$$

which are seen to be equal to the values given in 4(44) and 4(45).

For a transparent film on a transparent substrate, r_1, r_2 and δ_1 are all real. The reflectance and transmittance may be written

$$R = \frac{r_1^2 + 2r_1r_2 \cos 2\delta_1 + r_2^2}{1 + 2r_1r_2 \cos 2\delta_1 + r_1^2 r_2^2} \qquad \text{.... 4(109a)}$$

$$T = \frac{n_2}{n_0} \cdot \frac{t_1^2 t_2^2}{1 + 2r_1r_2 \cos 2\delta_1 + r_1^2 r_2^2} \qquad \text{.... 4(109b)}$$

and these forms are useful for this case. When film or substrate or both are absorbing, these forms are of no service and the expressions given in Section 4.9 below should be used.

Recurrence relations

The expressions for the reflectance and transmittance of a system of n films may be readily extended to deal with the case of an added film. If a $(n+1)^{th}$ film is added, then the Fresnel coefficients r_{n+1}, t_{n+1} in the system of n films must be replaced by the Fresnel coefficients r'_{n+1}, t'_{n+1} for the added film. From equations 4(108) these are seen to be given by

$$r'_{n+1} = \frac{r_{n+1}e^{i\delta_{n+1}} + r_{n+2}e^{-i\delta_{n+1}}}{e^{i\delta_{n+1}} + r_{n+1}r_{n+2}e^{-i\delta_{n+1}}} \qquad \text{.... 4(110a)}$$

$$t'_{n+1} = \frac{t_{n+1}t_{n+2}}{e^{i\delta_{n+1}} + r_{n+1}r_{n+2}e^{-i\delta_{n+1}}} \qquad \text{.... 4(110b)}$$

Thus in evaluating the matrix product 4(103), the final matrix is replaced by

$$\begin{pmatrix} e^{i\delta_n} & r'_{n+1}e^{i\delta_n} \\ r'_{n+1}e^{-i\delta_n} & e^{-i\delta_n} \end{pmatrix}$$

and the t_{n+1} in 4(104) is replaced by t'_{n+1}.

Since the matrix product is evaluated in a stepwise fashion, the calculation of the effect of an extra layer placed at the end of the pile is not unduly laborious.

Terse though the forms 4(106a) and 4(106b), the labour involved in the evaluation is considerable. The elements of the product matrix $\begin{pmatrix} a & b \\ c & d \end{pmatrix}$ are obtained by the usual rules for matrix multiplication. From 4(102) it is seen that the matrix elements are complex. For a system of transparent layers, the Fresnel coefficients are real and the matrices Hermitean. For an absorbing layer, the Fresnel coefficients for each boundary of the layer are complex and the resulting matrices non-Hermitean. The case of systems of absorbing films at non-normal incidence, where the angles involved in the expressions for the Fresnel coefficients are also complex, is depressing indeed.

4.9. Application of Matrix Method for Evaluating Reflectance and Transmittance

Explicit, single expressions for the reflectance and transmittance of systems of many films are cumbersome and of no great use. Since the computation of such expressions normally proceeds in a step-wise fashion, there is no disadvantage in giving the expressions for **R** and **T** in the form of successive stages of equations. The equations for systems of many layers are long and numerous. Results will therefore be given only for a few special cases which are likely to be of frequent service. They will serve to indicate the general manner in which the calculations may be performed and will establish a simple notation which may be conveniently applied to the problem of any number of layers. The special case of a system in which a film combination is repeated many times has been dealt with by L. I. Epstein[18].

The results will be quoted for normal incidence only. The extension of the treatment to the case of non-normal incidence may be easily effected for homogeneous isotropic layers, albeit at some length.

We first introduce a notation for the matrix elements which will enable expressions relating to any number of layers to be written in a form which is both succinct and convenient for numerical computation. It was shown above that the results are conveniently expressed in terms of Fresnel coefficients and these are, in general, complex. For the coefficients at the mth interface we write $r_m = g_m + ih_m$ and $t_m = 1 + g_m + ih_m$. If both the $(m-1)^{th}$ and the mth media are absorbing, with $\boldsymbol{n}_{m-1} = n_{m-1} - ik_{m-1}$ and $\boldsymbol{n}_m = n_m - ik_m$, then for the case of normal incidence

$$g_m = \frac{n_{m-1}^2 + k_{m-1}^2 - n_m^2 - k_m^2}{(n_{m-1} + n_m)^2 + (k_{m-1} + k_m)^2},$$

$$h_m = \frac{2(n_{m-1}k_m - n_m k_{m-1})}{(n_{m-1} + n_m)^2 + (k_{m-1} + k_m)^2} \qquad \text{.... } 4(111)$$

If the metrical thickness of the $(m-1)^{th}$ layer is d_{m-1} then the phase term $e^{i\delta_{m-1}}$ in the mth matrix is written

$$\exp i\delta_{m-1} = \exp \frac{i2\pi}{\lambda}(n_{m-1} - ik_{m-1})d_{m-1}$$

$$= \exp \alpha_{m-1} \exp i\gamma_{m-1} \qquad \text{.... } 4(112)$$

74

where $\quad \alpha_{m-1} = \dfrac{2\pi}{\lambda} k_{m-1} d_{m-1} \quad$ and $\quad \gamma_{m-1} = \dfrac{2\pi}{\lambda} n_{m-1} d_{m-1}.$

We note that all the elements of the matrices are complex and we write the mth matrix as

$$(C_m) = \begin{pmatrix} p_m + iq_m & r_m + is_m \\ t_m + iu_m & v_m + iw_m \end{pmatrix}$$

A double suffix notation serves to denote the elements of product matrices. Thus the elements of $(C_1)(C_2)$ are written

$$\begin{pmatrix} p_{12} + iq_{12} & r_{12} + is_{12} \\ t_{12} + iu_{12} & v_{12} + iw_{12} \end{pmatrix}$$

The elements of (C_1) (C_2) (C_3) are written

$$\begin{pmatrix} p_{13} + iq_{13} & r_{13} + is_{13} \\ t_{13} + iu_{13} & v_{13} + iw_{13} \end{pmatrix}$$

This form may be used unambiguously for $(C_1)(C_2)(C_3)$ since the product $(C_1)(C_3)$, with which it might ordinarily be confused, does not occur in multi-layer theory.

The rule for multiplying matrices leads to the following recurrence relation enabling the elements of $(C_1)(C_2) \ldots (C_{n-1})$ to be obtained.

$$\left.\begin{aligned}
p_{1,n+1} &= p_{1n}p_n - q_{1n}q_n + r_{1n}t_n - s_{1n}u_n \\
q_{1,n+1} &= q_{1n}p_n + p_{1n}q_n + s_{1n}t_n + r_{1n}u_n \\
r_{1,n+1} &= p_{1n}r_n - q_{1n}s_n + r_{1n}v_n - s_{1n}w_n \\
s_{1,n+1} &= q_{1n}r_n + p_{1n}s_n + s_{1n}v_n + r_{1n}w_n \\
t_{1,n+1} &= t_{1n}p_n - u_{1n}q_n + v_{1n}t_n - w_{1n}u_n \\
u_{1,n+1} &= u_{1n}p_n + t_{1n}q_n + w_{1n}t_n + v_{1n}u_n \\
v_{1,n+1} &= t_{1n}r_n - u_{1n}s_n + v_{1n}v_n - w_{1n}w_n \\
w_{1,n+1} &= u_{1n}r_n + t_{1n}s_n + w_{1n}v_n + v_{1n}w_n
\end{aligned}\right\} \quad \ldots\ 4(113)$$

The matrix elements are readily found from 4(102) and 4(111–2)

$$(C_m) = \begin{pmatrix} e^{i\delta_{m-1}} & r_m e^{i\delta_{m-1}} \\ r_m e^{-i\delta_{m-1}} & e^{-i\delta_{m-1}} \end{pmatrix} \equiv \begin{pmatrix} p_m + iq_m & r_m + is_m \\ t_m + iu_m & v_m + iw_m \end{pmatrix}$$

and

$$p_m = e^{\alpha_{m-1}} \cos \gamma_{m-1}$$
$$q_m = e^{\alpha_{m-1}} \sin \gamma_{m-1}$$
$$r_m = e^{\alpha_{m-1}} (g_m \cos \gamma_{m-1} - h_m \sin \gamma_{m-1})$$

$$s_m = e^{a_{m-1}}\left(h_m \cos \gamma_{m-1} + g_m \sin \gamma_{m-1}\right)$$

$$t_m = e^{-a_{m-1}}\left(g_m \cos \gamma_{m-1} + h_m \sin \gamma_{m-1}\right)$$

$$u_m = e^{-a_{m-1}}\left(h_m \cos \gamma_{m-1} - g_m \sin \gamma_{m-1}\right)$$

$$v_m = e^{-a_{m-1}} \cos \gamma_{m-1}$$

$$w_m = -e^{-a_{m-1}} \sin \gamma_{m-1}$$

(a) *The single film*

(i) *Single absorbing layer on an absorbing substrate. (Figure 4.9.)*—Some simplification arises in applying the general equations above since the first matrix is $\begin{pmatrix} 1 & g_1 + ih_1 \\ g_1 + ih_1 & 1 \end{pmatrix}$, of simpler form than the general term.

For a layer of index $\boldsymbol{n}_1 \equiv n_1 - ik_1$ and thickness d_1 on a substrate of

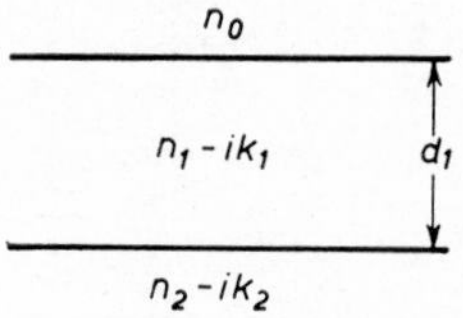

Figure 4.9   Absorbing film on an absorbing substrate

index $\boldsymbol{n}_2 \equiv n_2 - ik_2$, the reflectance at a wavelength λ may be found from the following equations

$$g_1 = \frac{n_0^2 - n_1^2 - k_1^2}{(n_0 + n_1)^2 + k_1^2} \qquad h_1 = \frac{2n_0 k_1}{(n_0 + n_1)^2 + k_1^2}$$

$$g_2 = \frac{n_1^2 - n_2^2 + k_1^2 - k_2^2}{(n_1 + n_2)^2 + (k_1 + k_2)^2} \qquad h_2 = \frac{2(n_1 k_2 - n_2 k_1)}{(n_1 + n_2)^2 + (k_1 + k_2)^2}$$

$$p_2 = e^{a_1} \cos \gamma_1 \qquad q_2 = e^{a_1} \sin \gamma_1$$

$$t_2 = e^{-a_1}(g_2 \cos \gamma_1 + h_2 \sin \gamma_1)$$

$$u_2 = e^{-a_1}(h_2 \cos \gamma_1 - g_2 \sin \gamma_1)$$

where $\quad a_1 = \dfrac{2\pi k_1 d_1}{\lambda} \quad$ and $\quad \gamma_1 = \dfrac{2\pi n_1 d_1}{\lambda} \quad$ (radians)

$$p_{12} = p_2 + g_1 t_2 - h_1 u_2$$

$$q_{12} = q_2 + h_1 t_2 + g_1 u_2$$

$$t_{12} = t_2 + g_1 p_2 - h_1 q_2$$

$$u_{12} = u_2 + h_1 p_2 + g_1 q_2$$

76

Then
$$R_1 = \frac{t_{12}^2 + u_{12}^2}{p_{12}^2 + q_{12}^2} \qquad \text{.... } 4(114)$$

The equations given in this form are of use in extending the treatment to deal with more than one film. For the single film case, however, some simplification can be effected and the expression for the reflectance given in the following form

$$R_1 = \frac{(g_1^2 + h_1^2)e^{2a_1} + (g_2^2 + h_2^2)e^{-2a_1} + A\cos 2\gamma_1 + B\sin 2\gamma_1}{e^{2a_1} + (g_1^2 + h_1^2)(g_2^2 + h_2^2)e^{-2a_1} + C\cos 2\gamma_1 + D\sin 2\gamma_1} \qquad \text{.... } 4(115a)$$

where
$$A = 2(g_1 g_2 + h_1 h_2) \qquad\qquad B = 2(g_1 h_2 - g_2 h_1)$$
$$C = 2(g_1 g_2 - h_1 h_2) \qquad\qquad D = 2(g_1 h_2 + g_2 h_1)$$

We may note here that, in contrast to the case of the transparent film, for which the reflectance at the air side and that of the substrate side are equal, the reflectances on each side of an absorbing layer are in general different. Replacing r_1 by $-r_2$ and r_2 by $-r_1$ in the derivation of the above equation, we obtain

$$R_1' = \frac{(g_2^2 + h_2^2)e^{2a_1} + (g_1^2 + h_1^2)e^{-2a_1} + A\cos 2\gamma_1 + B\sin 2\gamma_1}{e^{2a_1} + (g_1^2 + h_1^2)(g_2^2 + h_2^2)e^{-2a_1} + C\cos 2\gamma_1 + D\sin 2\gamma_1} \qquad \text{.... } 4(115b)$$

(ii) *Single absorbing layer on a transparent substrate.*—For a transparent substrate, $k_2 = 0$ and slight simplification results for the values of g_2 and h_2.

For a layer of this type, the transmittance of the film may usefully be calculated. Moreover, calculation of this quantity serves as a check on the arithmetic since $R + T$ cannot exceed unity. With heavily absorbing layers, of thickness such that multiple reflections may be neglected, the value obtained for the transmittance may be checked against the approximate value $\dfrac{n_2}{n_0}\exp\left(\dfrac{-4\pi k_1 d_1}{\lambda}\right)$

$$T_1 = \frac{n_2}{n_0} \cdot \frac{[(1 + g_1)^2 + h_1^2][(1 + g_2)^2 + h_2^2]}{e^{2a_1} + (g_1^2 + h_1^2)(g_2^2 + h_2^2)e^{-2a_1} + C\cos 2\gamma_1 + D\sin 2\gamma_1} \qquad \text{.... } 4(116)$$

It is sometimes of interest to know the reflectance of a film on the substrate side, when the film is deposited on a transparent substrate. This is obtained from the equations given in section (i) above where k_2 is put equal to zero, the substrate index inserted for n_0 and n_2 used for the index of air.

The reflectance of the film is in general different on the two sides, when the bounding regions are of different refractive index. The transmittance is the same for both directions of travel of the light.

77

(iii) *Transparent film on a transparent substrate.*—Considerable simplification results. We have $h_1 = h_2 = 0$, $a_1 = 0$, $A = C = 2g_1g_2$ and $B = D = 0$. The expressions for R and T reduce to

$$R_1 = \frac{g_1^2 + g_2^2 + 2g_1g_2 \cos 2\gamma_1}{1 + g_1^2g_2^2 + 2g_1g_2 \cos 2\gamma_1} \qquad \text{.... 4(117a)}$$

$$T_1 = \frac{n_2}{n_0} \cdot \frac{(1+g_1)^2(1+g_2)^2}{1 + g_1^2g_2^2 + 2g_1g_2 \cos 2\gamma_1} \qquad \text{.... 4(117b)}$$

These are seen to agree with equations 4(48)–4(49) since, for the transparent film on transparent substrate, $g_1 = r_1$ and $g_2 = r_2$.

Finally, for a transparent layer of refractive index n_1 immersed in a medium of index n_0, further reduction leads to

$$R_1 = \frac{4g_1^2 \sin^2 \gamma_1}{1 + g_1^4 - 2g_1^2 \cos 2\gamma_1} \qquad \text{.... 4(118a)}$$

$$T_1 = \frac{(1 - g_1^2)^2}{1 + g_1^4 - 2g_1^2 \cos 2\gamma_1} \qquad \text{.... 4(118b)}$$

(b) *The double layer*

(i) *Two absorbing layers on an absorbing substrate.*—The above treatment of a single absorbing layer on an absorbing substrate is readily extended to the case of two layers, shown in *Figure 4.10*. In addition to the quantities given under (a)(i) we require

$$g_3 = \frac{n_2^2 - n_3^2 + k_2^2 - k_3^2}{(n_2 + n_3)^2 + (k_2 + k_3)^2} \qquad h_3 = \frac{2(n_2k_3 - n_3k_2)}{(n_2 + n_3)^2 + (k_2 + k_3)^2}$$

$$a_2 = \frac{2\pi}{\lambda} k_2d_2 \qquad \gamma_2 = \frac{2\pi}{\lambda} n_2d_2 \quad \text{(radians)}$$

$$p_3 = e^{a_2} \cos \gamma_2 \qquad q_3 = e^{a_2} \sin \gamma_2$$

$$t_3 = e^{-a_2}(g_3 \cos \gamma_2 + h_3 \sin \gamma_2)$$

$$u_3 = e^{-a_2}(h_3 \cos \gamma_2 - g_3 \sin \gamma_2)$$

$$r_2 = e^{a_1}(g_2 \cos \gamma_1 - h_2 \sin \gamma_1)$$

$$s_2 = e^{a_1}(h_2 \cos \gamma_1 + g_2 \sin \gamma_1)$$

$$v_2 = e^{-a_1} \cos \gamma_1 \qquad w_2 = -e^{-a_1} \sin \gamma_1$$

$$r_{12} = r_2 + g_1v_2 - h_1w_2$$

$$s_{12} = s_2 + h_1v_2 + g_1w_2$$

$$v_{12} = v_2 + g_1r_2 - h_1s_2$$

$$w_{12} = w_2 + h_1r_2 + g_1s_2$$

$$p_{13} = p_{12}p_3 - q_{12}q_3 + r_{12}t_3 - s_{12}u_3$$
$$q_{13} = q_{12}p_3 + p_{12}q_3 + s_{12}t_3 + r_{12}u_3$$
$$t_{13} = t_{12}p_3 - u_{12}q_3 + v_{12}t_3 - w_{12}u_3$$
$$u_{13} = u_{12}p_3 + t_{12}q_3 + w_{12}t_3 + v_{12}u_3$$

whence
$$\boldsymbol{R}_2 = \frac{t_{13}^2 + u_{13}^2}{p_{13}^2 + q_{13}^2} \qquad \dots \ 4(119)$$

If the substrate be non-absorbing and the transmittance of the system is required, we need, in addition

$$l_{13} = (1 + g_1)(1 + g_2)(1 + g_3) - h_2 h_3 (1 + g_1) - h_3 h_1 (1 + g_2) - h_1 h_2 (1 + g_3)$$
$$m_{13} = h_1 (1 + g_2)(1 + g_3) + h_2 (1 + g_3)(1 + g_1) + h_3 (1 + g_1)(1 + g_2) - h_1 h_2 h_3$$

and
$$\boldsymbol{T}_2 = \frac{n_3}{n_0} \frac{l_{13}^2 + m_{13}^2}{p_{13}^2 + q_{13}^2} \qquad \dots \ 4(120)$$

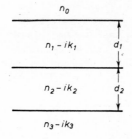

Figure 4.10 *Two absorbing films on an absorbing substrate*

As an arithmetical check during the computation for this system, the values of $\boldsymbol{R}_1$ and $\boldsymbol{T}_1$, given by 4(114) and 4(116) should be calculated at that stage. A sensible result indicates that the calculations may be correct. An error is indicated with complete certainty if the first layer prove to be a light-generator with $\boldsymbol{R}_1 + \boldsymbol{T}_1 > 1$.

(ii) *Two transparent layers on a transparent substrate.*—Putting $k_1 = k_2 = k_3 = 0$, we obtain

$$g_1 = \frac{n_0 - n_1}{n_0 + n_1} \qquad\qquad g_2 = \frac{n_1 - n_2}{n_1 + n_2} \qquad\qquad g_3 = \frac{n_2 - n_3}{n_2 + n_3}$$

With $\gamma_1 = \dfrac{2\pi}{\lambda} n_1 d_1$ and $\gamma_2 = \dfrac{2\pi}{\lambda} n_2 d_2$ (both in radians), we find

$$p_{13} = (1 + g_1 g_2 + g_2 g_3 + g_3 g_1) \cos \gamma_1 \cos \gamma_2 -$$
$$(1 - g_1 g_2 + g_2 g_3 - g_3 g_1) \sin \gamma_1 \sin \gamma_2$$

$$q_{13} = (1 - g_1g_2 - g_2g_3 + g_3g_1) \sin \gamma_1 \cos \gamma_2 +$$
$$(1 + g_1g_2 - g_2g_3 - g_3g_1) \cos \gamma_1 \sin \gamma_2$$

$$t_{13} = (g_1 + g_2 + g_3 + g_1g_2g_3) \cos \gamma_1 \cos \gamma_2 -$$
$$(g_1 - g_2 + g_3 - g_1g_2g_3) \sin \gamma_1 \sin \gamma_2$$

$$u_{13} = (g_1 - g_2 - g_3 + g_1g_2g_3) \sin \gamma_1 \cos \gamma_2 +$$
$$(g_1 + g_2 - g_3 - g_1g_2g_3) \cos \gamma_1 \sin \gamma_2$$

$$l_{13} = (1 + g_1)(1 + g_2)(1 + g_3)$$

Then
$$R_2 = \frac{t_{13}^2 + u_{13}^2}{p_{13}^2 + q_{13}^2}$$

and
$$T_2 = \frac{n_3}{n_0} \cdot \frac{l_{13}^2}{p_{13}^2 + q_{13}^2}$$

$$\left. \right\} \quad \ldots \ 4(121)$$

4.10. GRAPHICAL METHODS
(APPROXIMATE FOR TRANSPARENT LAYERS)

Since the amplitude and phase of a light beam may be conveniently represented by a vector, the properties of systems of transparent layers may be represented by graphical constructions. Very simple constructions may be used for systems of layers in which the Fresnel coefficients are small enough for multiple reflections to be neglected. The range of conditions under which the error introduced is negligible is small but is such as to include many cases of practical interest.

Reflection at a single transparent film on a transparent substrate

If we neglect multiple reflections within the film, then we regard the reflected beam as made up simply of two beams (*Figure 4.11*) whose amplitudes are given, for an incident beam of unit amplitude, by

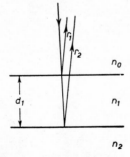

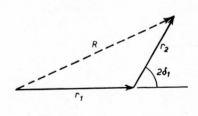

Figure 4.11 Reflections at a single film (neglecting multiple reflections within the film)

Figure 4.12 Vector diagram corresponding to single film

the Fresnel reflection coefficients r_1, r_2 and with phase difference determined by the film thickness. The resultant (*Figure 4.12*) is the vector sum of r_1, r_2 with phase difference $2\delta_1 = \dfrac{4\pi}{\lambda} n_1 d_1$ due to the double path through the film. The condition for an anti-reflecting film is readily seen to be that in which $r_1 = r_2$ and $2\delta_1 = \pi$ leading to the well-known relations that $n_1^2 = n_0 n_2$ and $n_1 d_1 = \lambda/4$. If this condition is met for a wavelength in the centre of the visible spectrum, then the vector diagrams for wavelengths at the two ends of the spectrum are shown in *Figure 4.13(a)* and *(c)*.

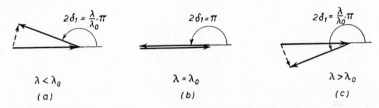

$$2\delta_1 = \frac{\lambda}{\lambda_0}.\pi \qquad\qquad 2\delta_1 = \pi \qquad\qquad 2\delta_1 = \frac{\lambda}{\lambda_0}.\pi$$

$$\lambda < \lambda_0 \qquad\qquad\quad \lambda = \lambda_0 \qquad\qquad\quad \lambda > \lambda_0$$

$$(a) \qquad\qquad\qquad (b) \qquad\qquad\qquad (c)$$

Figure 4.13   Vector diagram corresponding to a single anti-reflecting film for wavelengths around λ_0

The error introduced by the neglect of multiple reflections is easily deduced. From *Figure 4.12* we see that the resultant reflected energy is given by

$$R_{\text{approx}} = r_1^2 + r_2^2 + 2r_1 r_2 \cos 2\delta_1 \qquad \text{.... } 4(122)$$

The exact result is given by

$$R = \frac{r_1^2 + r_2^2 + 2r_1 r_2 \cos 2\delta_1}{1 + 2r_1 r_2 \cos 2\delta_1 + r_1^2 r_2^2} \qquad \text{.... } 4(123)$$

as shown in Section 4.4.

The two values differ by the factor $(1 + 2r_1 r_2 \cos 2\delta_1 + r_1^2 r_2^2)$ in the denominator of equation 4(123) which is seen to contain only

Table 4.1

Refractive Index of Layer (n_1)	R_{approx} (neglecting multiple reflections)	R (exact)
1·2	0·0399	0·0392
1·4	0·0409	0·0404
1·7	0·0387	0·0399
2·0	0·0364	0·0456
2·5	0·032	0·0460

second and higher powers of the Fresnel coefficients. When r_1 and r_2 are $\ll 1$, we should expect a reasonably close value to be given by the graphical method. An idea of the range of index for which the error is small is obtained from the table above which is computed for a half-wavelength film of optical thickness $\lambda/2$, where $\lambda = 5600$ Å, on a substrate of refractive index $1{\cdot}50$. $n_0 = 1$.

For indices ranging up to about $1{\cdot}7$, the error is less than about 2 per cent, which is often of no dire consequence. At higher indices, for which large Fresnel coefficients apply, the error becomes considerable.

Multiple layers

The exact calculation of reflectance for a single, transparent film is easy and involves but little labour. The vector method comes into its own for systems of two or more layers, for which the computations become laborious. Thus for the two-layer system in *Figure 4.14*, the vector diagrams for several wavelengths are shown in *Figure 4.15*. The Fresnel coefficients are the same in all cases, if the dispersion of the films is negligible over the range of wavelength considered, and the diagrams for different wavelengths are made by simply giving the vectors the appropriate orientations, calculated from the known optical thicknesses.

The variation of reflectance with angle of incidence may be dealt with by this method, the appropriate values of the Fresnel coefficients being obtained from equations 4(19)–4(22).

We defer until Chapter 7 the detailed consideration of high-reflecting and low-reflecting systems of films, merely remarking in passing that, for a large resultant reflectance, the thickness of the layers must be chosen so that the vectors representing the Fresnel coefficients lie in the same straight line and have the same sense; whilst for an anti-reflecting system, the vector polygon must close. Thus for a substrate bearing a low-index layer, followed by a high-index layer (in that order) of the same optical thickness $\lambda_0/4$, the vector diagram for a wavelength λ_0 is shown in *Figure 4.16* whilst the case of a substrate bearing a high-index layer of thickness $\lambda_0/2$ followed by a low-index layer of thickness $\lambda_0/4$ gives a vector diagram like that of *Figure 4.17* by judicious choice of refractive indices.

4.11. Graphical Methods (General)

Various workers (references [12–14]) have introduced graphical methods in which no approximation is involved for dealing with the problem of finding the reflectances and transmittances of

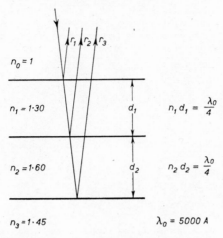

Figure 4.14   *Double layer system, anti-reflecting at* λ_0

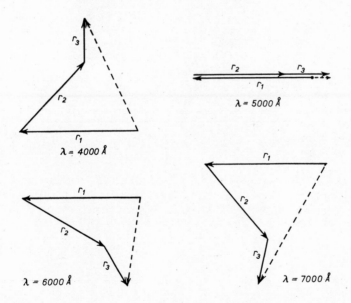

Figure 4.15   *Vector diagrams for a double film giving reflected amplitudes at wavelengths around* λ_0

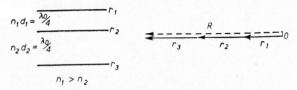

Figure 4.16   The low-index high-index quarter-wave pair, with the vector diagram for $\lambda = \lambda_0$

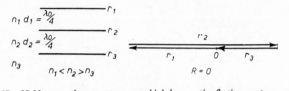

Figure 4.17   Half-wave plus quarter-wave high-low anti-reflecting system with vector diagram

systems of films.   After the slight initial labour of constructing suitable nomograms, these methods enable results to be obtained easily to within about 1 per cent, an accuracy quite sufficient for many purposes, and obtained with incomparably less labour than is the case if direct calculation is employed.

D. MALÉ[15] uses a nomogram for determining the modulus and argument of the complex Fresnel coefficients, together with a graphical construction from which the reflectance and transmittance of a film may be obtained.   The scheme may be applied very simply to the case of a single absorbing film on a transparent substrate or, with slight additional complication, to the problem of a film on an absorbing substrate.   We shall see that by the use of Rouard's scheme (Section 4.6) the method may be extended to the case of multiple layers.

Considering an absorbing film, of index $n_1 - ik_1$ on a transparent substrate of index n_2, the Fresnel coefficients for light travelling into the film from a medium of index n_0 may be written

$$r_1 = -\frac{\dfrac{n_1 - ik_1}{n_0} - 1}{\dfrac{n_1 - ik_1}{n_0} + 1} = -\sigma_1 e^{-i\beta_1} \qquad \text{.... } 4(124a)$$

$$r_2 = \frac{\dfrac{n_1 - ik_1}{n_2} - 1}{\dfrac{n_1 - ik_1}{n_2} + 1} = \sigma_2 e^{-i\beta_2} \qquad \text{.... } 4(124b)$$

84

each of which is of the form $\sigma e^{-i\beta} = \pm \dfrac{z-1}{z+1}$. The loci of constant β and of constant σ form orthogonal sets of circles so that a nomogram enabling β and σ to be read directly from given values of $z \equiv \dfrac{n_1}{n_0} - i\dfrac{k_1}{n_0}$ is easily constructed. Such a nomogram for values of n_1/n_0 from 0 to 5 and k_1/n_0 from 0 to 6 is shown in *Figure 4.18*. (The

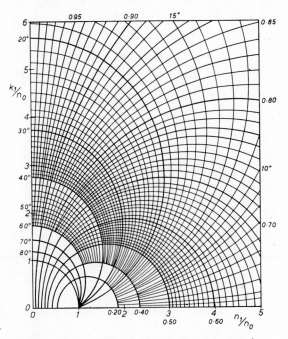

Figure 4.18   Malé's nomogram for determining r_1 and r_2

angles marked on the scales are the positive values of β. Thus for a film of refractive index $n = 2$ and extinction coefficient $k = 1$ in air $(n_0 = n_2 = 1)$ the values of r_1, r_2 are

$$r_1 = -0 \cdot 45 \ \text{cis} \ (-26 \cdot 5°)$$
$$r_2 = +0 \cdot 45 \ \text{cis} \ (-26 \cdot 5°) \ )$$

The amplitude reflection coefficient for a film of thickness d and for light of wavelength λ is given by equation 4(68) which may be written

$$\frac{E_0^-}{E_0^+} = \frac{1}{r_1} \cdot \frac{r_1/r_2 + e^{-2i\delta_1}}{1/r_1 r_2 + e^{-2i\delta_1}} = \frac{1}{r_1} \cdot P e^{i\Theta} \qquad \text{.... } 4(125)$$

Determination of P enables the reflectance, $\boldsymbol{R}$, and the phase change on reflection, Θ, to be calculated. These quantities are easily found from a simply-constructed vector diagram. Since r_1 and r_2 have already been found in the form $\sigma e^{-i\beta}$, the values of r_1/r_2 $(\mathcal{N})$

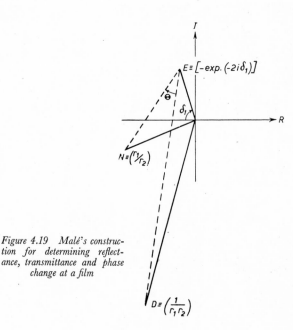

Figure 4.19 Malé's construction for determining reflectance, transmittance and phase change at a film

and $1/r_1r_2$ (D) are readily found and plotted on an Argand diagram (*Figure 4.19*). The point $E \equiv -\exp(-2i\delta_1) = -\exp\left(-\dfrac{4\pi k_1 d}{\lambda}\right)\exp\left(\dfrac{-4\pi i n_1 d}{\lambda}\right)$ is plotted, and we have $P = EN/ED$ and $\Theta = \widehat{NED}$.

The reflectance is given by $\left|\dfrac{E_0^-}{E_0^+}\right|^2 = \dfrac{P^2}{\sigma_1^2}$, where σ_1 is the modulus of r_1 and the phase change on reflection is given by $(\Theta - \beta_1)$.

Thus for a film for which $d/\lambda = 0.05$ and with optical constants $n_1 = 2$ and $k_1 = 1$, on a substrate of index $n_2 = 1.50$, we find, from the nomogram (*Figure 4.18*)

$$r_1 = -0.45 \;\underline{|-26.5°} \qquad\qquad r_2 = 0.30_5 \;\underline{|-47.5°} \;\; .... \;\; 4(126)$$

Thus $\mathcal{N}$ is the point $-1.48_5 \,\underline{|21°}$, D the point $7.25\,\underline{|74°}$ and E is

$-0.534 \mid -72°$. From a graph similar to that shown in *Figure 4.19*, plotted on a scale 2 cm = 1 unit, we find $P = 0.202$ and the reflectance equal to 0.204 ± 0.005. The transmittance may be readily determined (see below) and is found to be 0.42_7. The computed values for such a film are $R = 0.208$ and $T = 0.430$. Similarly the values for the phase change on reflection at the air surface are $+6.5°$ (graphical) and $+6.8_2°$ (calculation).

The method is particularly suitable for investigating the variation of reflectance and phase change with thickness of a given film. The points D, N remain fixed whilst E winds up on the origin in the form of a logarithmic spiral, as d/λ (and therefore δ, equation 4(110)) increases. Malé has applied the method to this problem, evaluating the reflectances on the air and glass side of films with optical constants characteristic of metals. The occurrence of odd things like minima in the reflectance vs. thickness curve and abrupt changes in the phase change vs. thickness curve, for certain conditions, are seen to be the natural predictions of theory. Such experimental observations had in the past been thought to indicate anomalous behaviour of materials in the form of thin films. Although the thin film theory was at this time (1930–5) quite able to deal with the problem of thin, absorbing layers, the prohibitive labour involved in direct computing had meant that curves in sufficient number to enable the general form of reflectance curves to be appreciated had not been obtained.

The extension of the method to two or more layers follows closely the method employed by Rouard (Section 4.6). The modulus and argument of the Fresnel coefficient for the layer next to the substrate are obtained by the method given above. The second layer is then regarded as lying on a surface with this value of Fresnel coefficient. This latter value is then substituted for r_2 in equation 4(125) and the amplitude coefficient is determined by a repetition of the procedure given above. If there are but two layers, this gives the reflectance of the surface. If there are more layers, then this coefficient is again used in computing the amplitude ratio for the next layer, and so on.

Determination of the transmittance

The vector diagram constructed for the reflectance measurement also serves for the determination of the transmittance. Equation 4(46) may be written

$$\left| \frac{E_2^+}{E_0^+} \right| = \frac{t_1 t_2}{r_1 r_2} \cdot \frac{e^{-i\delta_1}}{1/r_1 r_2 + e^{-2i\delta_1}} \qquad \text{.... 4(127)}$$

87

We may thus write

$$T = \frac{n_2}{n_0}\left|\frac{E_2^+}{E_0^+}\right|^2 = K\frac{e^{-4\pi k_1 d/\lambda}}{|ED|^2}$$

where K is easily found by using the known transmittance of the film for $\delta_1 = 0$, viz. $4n_0 n_2/(n_0 + n_2)^2$. If E_0 is the point $(-1, 0)$, then

$$K = |E_0 D|^2 \cdot \frac{4n_0 n_2}{(n_0 + n_2)^2}$$

and

$$T = \frac{4n_0 n_2}{(n_0 + n_2)^2}\left|\frac{E_0 D}{ED}\right|^2 e^{-4\pi k_1 d/\lambda} \qquad \text{.... } 4(128)$$

The transparent layer on an absorbing substrate

The graphical method may be applied to the case of an absorbing layer on an absorbing substrate although for this case the nomogram

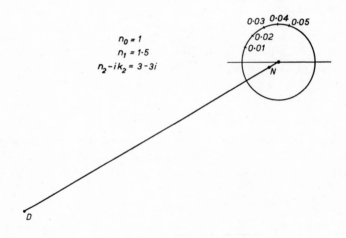

Figure 4.20 The vector diagram for a transparent film on an absorbing substrate. The marked points on the circle are the positions of E corresponding to the thicknesses stated (in fractions of a wavelength)

is of no service in obtaining the modulus and argument of the second Fresnel coefficient $r_2 = (n_1/n_2 - 1)/(n_1/n_2 + 1)$ since n_1 and n_2 are both complex. For a *transparent* layer on an absorbing substrate—which arrangement is found in certain types of interference filter—the nomogram may be used for the second Fresnel coefficient. For we may write

$$r_2 = -\frac{n_2/n_1 - 1}{n_2/n_1 + 1} \qquad \text{.... } 4(129)$$

where n_2 is complex and n_1 real. This is of the same form as 4(124a).

Furthermore, for a transparent layer, the logarithmic spiral, along which the point $E(\equiv -e^{-4\pi i n_1 d/\lambda})$ moves as the film thickness alters, degenerates into a circle. *Figure 4.20* illustrates the vector diagram obtained for the case of a film for which $d/\lambda = 0.01, 0.02$, etc. and having an index $n_1 = 1.50$ on a substrate for which $n_2 - ik_2 \equiv 3 - 3i$.

The transparent layer on a transparent substrate

The need for a nomogram disappears for this case, since both Fresnel coefficients are real. As in the case of a transparent layer on an absorbing base, the locus of E is a circle. The points N, D lie on the $\delta = 0$ axis.

The saving in time by using graphical methods of the type given above is very considerable—even when use is made of high-speed calculating devices—over that taken by direct computation. As P. COTTON[16] remarks on this aspect 'ces machines ne donnent pas cette vue d'ensemble des phénomènes qui est parfois plus utile qu'un calcul précis s'appliquant à un petit nombre de points.'

4.12. THE PHASE CHANGE ON REFLECTION AT OR TRANSMISSION THROUGH A THIN FILM

From the equations for the reflectance and transmittance of a single film (equations $4(44)$ and $4(45)$) we see that the reflected and transmitted amplitudes are complex. They may therefore be written as $R_0 = \varrho_0 e^{i\Delta_0}$ and $T_2 = \tau_2 e^{i\gamma_2}$ where ϱ_0, τ_2 are the real parts of the coefficients and Δ_0, γ_2 represent changes in phase on reflection at or transmission by the film. These phase changes may be measured and are of considerable importance, especially in connection with interferometric work employing metallized surfaces. The path equivalent of the phase change at the surface constitutes part of the optical path lengths occurring in the interferometer. The results of measurements of phase changes are given in Sections 6.5–7.

Transparent film

By separating the complex expressions for R_0, T_2 into real and imaginary parts, we may readily obtain the modulus and the phase

89

terms of the reflected and transmitted beams. Thus from 4(44) we have

$$R_0 = \varrho_0 e^{i\Delta_0} = \frac{r_1 + r_1 r_2^2 + (r_2 + r_1 r_2^2) \cos 2\delta_1}{1 + 2r_1 r_2 \cos 2\delta_1 + r_1^2 r_2^2} - i \cdot \frac{(r_2^2 - r_1^2 r_2) \sin 2\delta_1}{1 + 2r_1 r_2 \cos 2\delta_1 + r_1^2 r_2^2}$$

$$\text{.... } 4(130)$$

The single equation 4(130) represents both R_{0p}, R_{0s}. Values of r_1, r_2 are inserted appropriate to the plane of polarization under consideration.

For the transmitted beam we obtain from equation 4(45)

$$T_2 = \tau_2 e^{i\gamma_2} = \frac{t_1 t_2 (1 + r_1 r_2) \cos \delta_1}{1 + 2r_1 r_2 \cos 2\delta_1 + r_1^2 r_2^2} - i \frac{t_1 t_2 (1 - r_1 r_2) \sin \delta_1}{1 + 2r_1 r_2 \cos 2\delta_1 + r_1^2 r_2^2}$$

$$\text{.... } 4(131)$$

The phase changes Δ_0, γ_2 on reflection at or transmission by the film are thus given by

$$\tan \Delta_{0p,s} = \frac{r_2(1 - r_1^2) \sin 2\delta_1}{r_1(1 + r_2^2) + r_1^2(1 + r_1^2) \cos 2\delta_1} \quad \text{.... } 4(132)$$

and

$$\tan \gamma_{2p,s} = -\frac{(1 - r_1 r_2)}{(1 + r_1 r_2)} \tan \delta_1 \quad \text{.... } 4(133)$$

The quantities generally derived from polarimetric measurements are

$$\tan \psi = \frac{\varrho_{0p}}{\varrho_{0s}} \quad \text{and} \quad \Delta = \Delta_{0p} - \Delta_{0s} \quad \text{.... } 4(134)$$

and these are readily obtained from equations 4(130) to 4(133). It is convenient to write

$$a_1 \equiv r_{1p} + r_{2p} \cos 2\delta_1 \qquad\qquad b_1 \equiv r_{2p} \sin 2\delta_1$$
$$a_2 \equiv 1 + r_{1p} r_{2p} \cos 2\delta_1 \qquad\quad b_2 \equiv r_{1p} r_{2p} \sin 2\delta_1$$
$$a_3 \equiv r_{1s} + r_{2s} \cos 2\delta_1 \qquad\qquad b_3 \equiv r_{2s} \sin 2\delta_1$$
$$a_4 \equiv 1 + r_{1s} r_{2s} \cos 2\delta_1 \qquad\quad b_4 \equiv r_{1s} r_{2s} \sin 2\delta_1$$
$$A \equiv a_1 a_2 + b_1 b_2 \qquad\qquad\qquad B \equiv a_1 b_2 - a_2 b_1$$
$$A' \equiv a_3 a_4 + b_3 b_4 \qquad\qquad\qquad B' \equiv a_3 b_4 - a_4 b_3$$

We then have

$$\tan \psi = \frac{\{(AB' + A'B)^2 + (AA' - BB')^2\}^{\frac{1}{2}}}{(a_2^2 + b_2^2)(a_3^2 + b_3^2)} \quad \text{.... } 4(135)$$

and

$$\tan \Delta = \frac{AB' + A'B}{AA' - BB'} \quad \text{.... } 4(136)$$

Absorbing film

The expressions for the phase changes on reflection or transmission for an absorbing film may be obtained from equations 4(132) and 4(133) by substituting the appropriate complex values of the Fresnel coefficients. For the general case of non-normal incidence, the resulting expressions are miserably complicated and are of too little use to justify the space which they would occupy if written out. For normal incidence, reasonably tractable expressions for the phase changes may be obtained, conveniently from the reflectance expressed in the form

$$R_0 = \varrho_0 e^{i\varDelta_0} = \frac{(n_0 - \boldsymbol{n}_1)(\boldsymbol{n}_1 + n_2)e^{i\delta_1} + (n_0 + \boldsymbol{n}_1)(\boldsymbol{n}_1 - n_2)e^{-i\delta_1}}{(n_0 + \boldsymbol{n}_1)(\boldsymbol{n}_1 + n_2)e^{i\delta_1} + (n_0 - \boldsymbol{n}_1)(\boldsymbol{n}_1 - n_2)e^{-i\delta_1}} \quad \text{.... } 4(137)$$

where $\quad \delta_1 = \dfrac{2\pi}{\lambda}\boldsymbol{n}_1 d_1 = \dfrac{2\pi}{\lambda}(n_1 - ik_1)d_1$

and $\qquad e^{\pm i\delta_1} = \exp\left(\pm \dfrac{2\pi k_1 d_1}{\lambda}\right)\exp\left(\pm i\dfrac{2\pi n_1 d_1}{\lambda}\right)$

$$= e^{\pm K} \operatorname{cis}(\pm \mathcal{N})$$

We may write equation 4(137) in the form

$$R_0 = \frac{A + iB}{C + iD} \qquad \text{.... } 4(138)$$

where

$$
\left.
\begin{aligned}
A &= e^{K}\{[(n_0 - n_1)(n_1 + n_2) + k_1^2]\cos\mathcal{N} + k_1(n_0 - 2n_1 - n_2)\sin\mathcal{N}\} \\
&+ e^{-K}\{[(n_0 + n_1)(n_1 - n_2) - k_1^2]\cos\mathcal{N} - k_1(n_0 + 2n_1 - n_2)\sin\mathcal{N}\} \\
B &= e^{K}\{[(n_0 - n_1)(n_1 + n_2) + k_1^2]\sin\mathcal{N} - k_1(n_0 - 2n_1 - n_2)\cos\mathcal{N}\} \\
&- e^{-K}\{[(n_0 + n_1)(n_1 - n_2) - k_1^2]\sin\mathcal{N} + k_1(n_0 + 2n_1 - n_2)\cos\mathcal{N}\} \\
C &= e^{K}\{[(n_0 + n_1)(n_1 + n_2) - k_1^2]\cos\mathcal{N} + k_1(n_0 + 2n_1 + n_2)\sin\mathcal{N} \\
&+ e^{-K}\{[(n_0 - n_1)(n_1 - n_2) + k_1^2]\cos\mathcal{N} - k_1(n_0 - 2n_1 + n_2)\sin\mathcal{N}\} \\
D &= e^{K}\{[(n_0 + n_1)(n_1 + n_2) - k_1^2]\sin\mathcal{N} - k_1(n_0 + 2n_1 + n_2)\cos\mathcal{N}\} \\
&- e^{-K}\{[(n_0 - n_1)(n_1 - n_2) + k_1^2]\sin\mathcal{N} + k_1(n_0 - 2n_1 + n_2)\cos\mathcal{N}\}
\end{aligned}
\right\} 4(139)
$$

The phase change on reflection is then given by

$$\Delta_0 = \arctan\left(\frac{n_1 D + k_1 C}{k_1 D - n_1 C}\right) \qquad \text{.... } 4(140)$$

For the phase change on transmission at normal incidence through the film we have

$$T_2 = \tau_2 e^{i\gamma_2} = \frac{4n_0\boldsymbol{n}_1}{(n_0+\boldsymbol{n}_1)(\boldsymbol{n}_1+n_2)e^{i\delta_1}+(n_0-\boldsymbol{n}_1)(\boldsymbol{n}_1-n_2)e^{-i\delta_1}}$$

$$\left. = \frac{4n_0(n_1-ik_1)}{C+iD} \right\} \qquad 4(141)$$

so that

$$\gamma_2 = \arctan\left(\frac{n_1 D + k_1 C}{k_1 D - n_1 C}\right) \qquad \dots\ 4(142)$$

In Chapter 6 we discuss the results of measurements of the phase change on reflection at various metal films. Curves of Δ_0 vs. film thickness obtained experimentally are found to differ in no uncertain manner from the form calculated from the above theory into which values of (n_1, k_1) characteristic of the bulk metal are inserted.

4.13. General Theory of Doubly Refracting Films

We have dealt in the preceding part of this chapter with the case of optically isotropic films, both transparent and absorbing. A complete theory of the optical behaviour of a homogeneous film must, however, take account of the possibility of optical anisotropy of the film. Such a general problem is formidable, as can be readily imagined from the complexity of the general case of an isotropic homogeneous layer. The case of an optically anisotropic film (uniaxial or biaxial) in which one of the optic axes lies normal to the plane of the film has been dealt with by H. Schopper[17].

Consider light incident in medium 1 on an anisotropic, absorbing film of thickness d supported on an isotropic, non-absorbing medium 3. The plane of incidence is $x = 0$ and the angle of incidence ψ_0. (*Figure 4.21.*) The amplitudes of the beams in the three media are written in terms of wave vectors $\varkappa_0 \equiv \{\varkappa_{0x}, \varkappa_{0y}, \varkappa_{0z}\}$, $\varkappa_1 \equiv \{\varkappa_{1x}, \varkappa_{1y}, \varkappa_{1z}\}$ etc., where

$$\varkappa = \frac{2\pi\boldsymbol{n}}{\lambda} \qquad \dots\ 4(143)$$

(The magnetic permeability is assumed to be unity.)
For the electric vectors in the three media we have

$$\left.\begin{aligned}
E_0^+ &= (E_0^+)_0 \exp\{-i(\varkappa_{0y}y + \varkappa_{0z}z)\} \\
E_0^- &= (E_0^-)_0 \exp\{-i(\varkappa_{0x}x + \varkappa_{0y}y - \varkappa_{0z}z)\} \\
E_1^+ &= (E_1^+)_0 \exp\{-i(\varkappa_{1x}x + \varkappa_{1y}y + \varkappa_{1z}z)\} \\
E_1^- &= (E_1^-)_0 \exp\{-i(\varkappa_{1x}x + \varkappa_{1y}y - \varkappa_{1z}z)\} \\
E_2^+ &= (E_2^+)_0 \exp\{-i(\varkappa_{2x}x + \varkappa_{2y}y + \varkappa_{2z}z)\}
\end{aligned}\right\} \qquad \dots\ 4(144)$$

where two sets of equations exist corresponding to components polarized in or perpendicular to the plane of incidence. The index of the second medium is specified by three indices and three extinction coefficients.

$$\left.\begin{array}{l} \boldsymbol{n}_x = n_x - i k_x \\ \boldsymbol{n}_y = n_y - i k_y \\ \boldsymbol{n}_z = n_z - i k_z \end{array}\right\} \quad \dots \; 4(145)$$

The reflectances (at either side of the film) and the transmittances may be calculated in the way outlined in Section 4.4, by solution

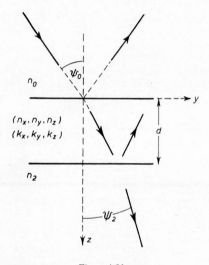

Figure 4.21

of Maxwell's equations subject to the usual boundary conditions at the film surfaces. The values of R, R' and T may be written in terms of Fresnel coefficients whose (complex and depressingly complicated) values may be derived in terms of the refractive indices, extinction coefficients and angles of incidence and emergence. It is convenient to define a_s, b_s and a_p, b_p by

$$\left.\begin{array}{l} a_s^2 - b_s^2 = n_x^2 - k_x^2 - n_0^2 \sin^2 \psi_0 \\ 2 a_s b_s = 2 n_x k_x \end{array}\right\} \quad \dots \; 4(146)$$

93

$$a_p^2 - b_p^2 = n_y^2 - k_y^2 - n_0^2 \sin^2 \psi_0 \cdot \frac{(n_y^2 - k_y^2)(n_z^2 - k_z^2) + 4n_y k_y n_z k_z}{(n_z^2 + k_z^2)^2}$$

$$2a_p b_p = 2n_y k_y + n_0^2 \sin^2 \psi_0 \cdot \frac{2n_z k_z (n_y^2 - k_y^2) - 2n_y k_y (n_z^2 - k_z^2)}{(n_z^2 + k_z^2)^2} \Bigg\} \quad \text{... } 4(147)$$

The Fresnel coefficients may then be simply expressed.

$$r_{1s} = \frac{n_0 \cos \psi_0 - a_s + ib_s}{n_0 \cos \psi_0 + a_s - ib_s} \qquad r_{2s} = \frac{a_s - ib_s - n_2 \cos \psi_2}{a_s - ib_s + n_2 \cos \psi_2} \Bigg\} \quad \text{.... } 4(148)$$

$$t_{1s} = 1 + r_{1s} \qquad\qquad t_{2s} = 1 + r_{2s}$$

and

$$r_{1p} = \frac{n_y^2 \cos \psi_0 - n_0(a_p - ib_p)}{n_y^2 \cos \psi_0 + n_0(a_p - ib_p)} \qquad r_{2p} = \frac{n_2(a_p - ib_p) - n_y^2 \cos \psi_2}{n_2(a_p - ib_p) + n_y^2 \cos \psi_2} \Bigg\} \quad 4(149)$$

$$t_{1p} = \frac{n_0}{n_y}(1 + r_{1p}) \qquad\qquad t_{2p} = \frac{n_y}{n_2}(1 + r_{2p})$$

The reflectances and transmittance are then readily though laboriously calculable from

$$R = \frac{r_1 + r_2 \exp\left[-4\pi i(a + ib)d/\lambda\right]}{1 + r_1 r_2 \exp\left[-4\pi i(a + ib)d/\lambda\right]} \qquad\qquad \text{.... } 4(150)$$

$$R' = \frac{r_2 + r_1 \exp\left[-4\pi i(a + ib)d/\lambda\right]}{1 + r_1 r_2 \exp\left[-4\pi i(a + ib)d/\lambda\right]} \exp\left(\frac{4\pi i n_2 d \cos \psi_2}{\lambda}\right) \quad 4(151)$$

$$T = \frac{t_1 t_2 \exp\left[-2\pi i(a + ib)d/\lambda\right]}{1 + r_1 r_2 \exp\left[-4\pi i(a + ib)d/\lambda\right]} \exp\left[(2\pi i n_2 d \cos \psi_2)/\lambda\right] \quad 4(152)$$

where the suffix p or s is affixed to give these quantities appropriate to reflection and transmission of light polarized with the electric vector respectively parallel or perpendicular to the plane of incidence.

Transparent anisotropic films

The expressions given above simplify considerably for non-absorbing films. The values of the Fresnel coefficients, which are now real, become

$$r_{1s} = \frac{n_0 \cos \psi_0 - (n_x^2 - n_0^2 \sin^2 \psi_0)^{\frac{1}{2}}}{n_0 \cos \psi_0 + (n_x^2 - n_0^2 \sin^2 \psi_0)^{\frac{1}{2}}}$$

$$r_{2s} = \frac{(n_x^2 - n_0^2 \sin^2 \psi_0)^{\frac{1}{2}} - n_2 \cos \psi_2}{(n_x^2 - n_0^2 \sin^2 \psi_0)^{\frac{1}{2}} + n_2 \cos \psi_2} \Bigg\} \quad \text{.... } 4(153)$$

and

$$r_{1p} = \frac{n_y \cos \psi_0 - n_0 \left[1 - \dfrac{n_0^2}{n_z^2} \sin^2 \psi_0\right]^{\frac{1}{2}}}{n_y \cos \psi_0 + n_0 \left[1 - \dfrac{n_0^2}{n_z^2} \sin^2 \psi_0\right]^{\frac{1}{2}}}$$

$$r_{2p} = \frac{n_2 \left[1 - \dfrac{n_0^2}{n_z^2} \sin^2 \psi_0\right]^{\frac{1}{2}} - n_y \cos \psi_2}{n_2 \left[1 - \dfrac{n_0^2}{n_z^2} \sin^2 \psi_0\right]^{\frac{1}{2}} + n_y \cos \psi_2}$$

$$t_{1p} = \frac{n_0}{n_y}(1 + r_{1p}) \qquad t_{2p} = \frac{n_y}{n_2}(1 + r_{2p})$$

.... 4(154)

The values of R, R' and T are then easily found from equations 4(150)–4(152) using the above values for the Fresnel coefficients. The a's and b's introduced above reduce to

$$a_s^2 = n_x^2 - n_0^2 \sin^2 \psi_0$$
$$a_p^2 = n_y^2 - \frac{n_0^2 n_y^2}{n_z^2} \sin^2 \psi_0 \qquad \text{Together with } b_s = b_p = 0$$

.... 4(155)

Transparent birefringent layers have been used in the frustrated total reflection filter (see Section 7.6), although no determinations of optical constants of such layers appear to have been made. Schopper[17] gives equations which enable the optical constants and thickness of such films to be determined.

REFERENCES

[1] ABELÈS, F.   *Ann. d. Physique* **3** (1948) 504
[2] MÜLLER-POUILLETS.   *Lehrbuch der Physik, II (Optik)*, Vieweg.
[3] BORN, M.   *Optik*, Springer
[4] *Handbuch der Physik* (1928), Vol. XX.   Springer
[5] DITCHBURN, R. W.   *Light*.   Blackie, 1952
[6] MAYER, H.   *Physik Dünner Schichte*.   Stuttgart, 1950
[7] CROOK, A. W.   *J.O.S.A.* **38** (1948) 954
[8] ROUARD, P.   *Ann. d. Physique* **7** (1937) 291
[9] VAŠÍČEK, A.   *J. d. Phys.* **11** (1950) 342
[10] BLAISSE, B. S.   *Ibid.* 315
[11] STRONG, J.   *Ibid.* 451
[12] PERROT, M. and COTTON, P.   *Ann. d. Physique* **20** (1945) 585
[13] WINTERBOTTOM, A. B.   *Trans. Far. Soc.* **42** (1946) 487
[14] POLSTER, H.   *J.O.S.A.* **36** (1946) 350
[15] MALÉ, D.   *J. d. Phys.* **11** (1950) 332
[16] COTTON, P.   *Ibid.* 375
[17] SCHOPPER, H.   *Z. Phys.* **132** (1952) 146
[18] EPSTEIN, L. I.   *J.O.S.A.* **42** (1952) 806

5

MEASUREMENTS OF FILM THICKNESS
AND OPTICAL CONSTANTS

The significance of film thickness—Mechanical methods for determining film thickness—Electrical methods—Optical methods—Multiple-beam interference methods—Measurement of the thickness and refractive index of transparent layers—Measurements of the thickness and optical constants of absorbing layers—The choice of method for the measurement of optical constants—Other optical methods for determining film thickness—Use of radioactive tracers

5.1. THE SIGNIFICANCE OF FILM THICKNESS

THE growth of the importance of thin films has been attended by considerable development of methods for determining film thickness and optical constants. The methods devised are many and varied and depend on such features as the nature of the film, whether transparent or absorbing, the thickness range which it is desired to cover and whether or not it is required to study changing thickness with time. A survey of methods which have been successfully employed is given in this chapter, with details of their ranges of usefulness and of their limitations.

Certain difficulties must, however, first be recognized relating to the concept of film thickness, especially in the case of films whose average thickness corresponds to a few atomic layers. The convenient picture which we may form of a film as an isotropic, homogeneous, plane parallel-sided layer is not in practice realized. Substrates on which films are formed are rarely mathematically plane. They are, furthermore, not inert in so far as their structures may exert a profound influence on the form of the overlying film. Moreover, for films prepared by sputtering or by vacuum evaporation surface motion of the condensed atoms is observed (Section 2.5), resulting in an aggregated, rather than a continuous, structure. For such a film we may define a mean thickness $\bar{t}$ by

$$\bar{t} = \frac{1}{A}\Sigma t\delta A \qquad \qquad \dots 5(1)$$

where A is the area of the film. This is the quantity which is obtained by certain of the methods given below. When the average diameter of the aggregates in a film is comparable with $\bar{t}$, then the mean thickness thus defined is of little useful significance; the mass per unit area is often used instead of the thickness. For a granular film, average thickness cannot be easily obtained from the mass per unit area, since the average density may differ widely from that of the bulk material. The thickness of film at which the bulk properties are attained varies considerably with the material. Thus although the optical properties of films of silver and aluminium are found to correspond very nearly with those of the bulk metals for thicknesses of a few hundred ångströms, those of films of alkali halides are detectably different from the bulk properties for thicknesses as high as several microns.

Of the methods to be described, most yield the average film thickness over an area of the order 1 cm². Certain of the optical methods, however, give the thickness at a point. The area of film covered by the 'point' is in these cases determined by the resolution limit of a microscope objective, and usually of a low-power one. The value of t obtained is thus the average over an area of the order of wavelengths in diameter. Since this is generally large compared with the scale of irregularities resulting from aggregation in the film, the value obtained is essentially the mean $\bar{t}$ as defined above.

The methods of measurement may be conveniently divided into three classes: (1) Mechanical Methods, (2) Electrical Methods, and (3) Optical Methods.

5.2. Mechanical Methods for Determining Film Thickness

(a) Weighing method

The sensitivity of many of the microbalances which have been designed in recent years is such as to bring within range of accurate mass-determination films of average thickness corresponding to only a small number of molecular layers. In the case of substances containing the heavier elements, a single molecular layer covering an area of the order of a square centimetre is within range of detection. If the mass of a parallel-sided film of area A and density ϱ is m, then the film thickness t is given simply by

$$t = \frac{m}{\varrho A} \qquad \dots 5(2)$$

Some mention has already been made of the difficulties of this method. The average density of very thin films (which is the appropriate value of ϱ for evaluating the average thickness) is known to be less than that of the bulk material on account of the granular structure of the film. The film density cannot be readily determined by other methods. Furthermore, the true film area A may not, in the event of the substrate departing violently from smoothness, correspond to its apparent area, as measured with a ruler. The true surface area may under these conditions exceed the apparent area by a considerable factor.

It is clearly of importance to know for what range of film thickness the weighing method may be relied upon to yield correct results and we may for this purpose be guided by the results obtained in experiments in which the thicknesses of weighed films are determined by an independent method. Thus P. L. CLEGG and A. W. CROOK[1], using a microbalance described below measured the thickness of weighed films of silver by Tolansky's multiple-beam interference method (Section 5.5) and found agreement to within a few per cent for films of thickness down to ~120 Å. Inspection of the electron micrographs of silver films given by R. S. SENNETT and G. D. SCOTT[2] shows that, for rapidly evaporated films (~100 Å/sec), the granular structure has nearly disappeared when thicknesses of the order 150 Å are reached. It seems, then, for such films that the density can differ but little from that of bulk silver. It should be noted, however, that films formed at a low rate of deposition (1–10 Å/min) show a much more marked granularity even at thicknesses of 400–500 Å, so that, for these films, the weighing method may be less reliable. The results of Clegg and Crook may be expected to be free from errors arising from the apparent surface effect since they used a mica substrate. The work of Tolansky has shown that mica crystal surfaces are locally smooth to within an atom layer.

The form of microbalance used by Clegg and Crook is shown in *Figure 5.1*. Films are deposited on the mica spade B, of area about 4 cm². The torsion wires are of phosphor-bronze strip, as used for galvanometer suspensions. The adjustable counterweights C and D ensure that the centre of gravity of the system can be made to lie on the axis of the torsion wire; this was found to be necessary in order to avoid drifts. Electrostatic shielding was also found to be required and the mica spade was covered with a metal film before assembly so that accumulating charges may leak away to the (metal) case of the balance via the suspension. Calibration with pieces of fine wire of known mass per unit length showed a typical sensitivity of 2×10^5 cm/gm, the deflection being measured by a lamp-and-scale at a distance of 1m. The weighings were found to be reliable for silver films of thickness down to ∼10 Å.

Details of the construction of an all-silica microbalance, suitable for use in the highest vacua and the theory of the action of the microbalance have been given by R. S. BRADLEY[3].

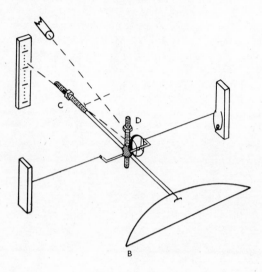

Figure 5.1   Microbalance designed by Clegg and Crook

(b) *Molecular ray method*

Under certain conditions it is possible to determine the mass of a film, prepared by the evaporation process, by measuring the source temperature and time of evaporation. The vapour pressure of the film material must be known and the film preparation must be carried out under conditions such that molecular streaming obtains. Moreover, in view of the behaviour of films observed by Estermann, Knudsen, *et al.* and described in Chapter 2, it must be established that all the incident atoms condense (accommodation coefficient unity). This entails the use of beam intensities higher than the critical values observed by Cockcroft and the use of targets at temperatures below the critical. Otherwise the phenomena of re-evaporation observed with very weak beams on substrates above the critical temperature may result in considerable inaccuracies in the film thickness determination by this method. Even when these conditions are complied with, the same uncertainties apply to the film *thickness* consequent upon the uncertainty in the density, as were mentioned in connection with the weighing method.

5.3. ELECTRICAL METHODS

Since electrical measurements may be made easily with high precision, it is not surprising that film thickness measurements have been attempted in which one or another electrical property has been used. Many of the measurements thus made have, when comparisons with non-electrical methods have been made, merely served to indicate the severely limited nature of this type of measurement. Gross differences have been observed between the electrical behaviour of matter in thin films and that of bulk materials so that measurements which rely on the use of bulk constants (resistivity, dielectric constant, etc.) are generally of little use.

(a) *Langmuir's ionization method*

Under certain limited conditions, discussed below, this early method may be applied to the determination of the thickness of films of certain alkali and alkaline earth elements. The method rests on the observation that when atoms of certain of these elements strike a heated tungsten wire (temperature $\sim 1000°$ C), they leave the metal surface as positive ions and not as neutral atoms. It was shown that, provided the ionization potential of the atom is smaller than the work function of the surface of the heated wire, all the atoms leaving the surface are charged. The use of a sensitive galvanometer, or latterly electrometer methods, for measuring the ionization current enables the strength of very weak beams of atoms to be determined.

Figure 5.2 shows schematically the arrangement used by I. LANGMUIR and K. H. KINGDON[4]. The alkali metal is boiled in oven O and after emerging from the orifice, some of the vapour strikes the heated filament W. All these atoms re-evaporate as ions and are swept to the collector electrode C by the applied field, the current I being measured in the process. The number of atoms striking W per unit time is then simply $N = i/e$ where e is the electronic change. If the heating current through W is then discontinued, keeping the oven temperature steady, the number of atoms condensing on the wire in time t (assuming that the beam intensity exceeds the critical value—see Section 2.5) is simply Nt.

Langmuir and Kingdon consider the problem of dealing with material which does not ionize completely on striking the wire. Work on this problem is, however, primarily in the nature of an investigation into processes of ionization, rather than a method for determining film thickness. A useful list of references to work of this type is to be found in Mayer's *Optik Dünner Schichte*.

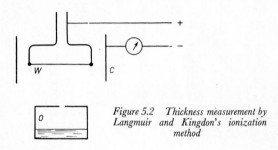

Figure 5.2   *Thickness measurement by Langmuir and Kingdon's ionization method*

(b) *Electrolytic methods*

At first sight, the determination of the mass of a layer deposited electrolytically appears straightforward and capable of high accuracy. The measurement of current and time, with the application of Faraday's laws, should suffice to determine the mass deposited. In certain investigations, in which the mass of deposit has been measured as a function of the time of deposition, Faraday's laws have been found to be obeyed. F. G. BROCKMAN's[5] investigations show this to be the case for nickel films of thickness up to 1000 Å and reveal the further interesting feature that the specific resistance of such films agrees with that obtained for the bulk material. This is in sharp contrast to the behaviour of films prepared by evaporation, and suggests a markedly different state of aggregation.

In many other cases, however, the method may prove unreliable. The deposit may form in a spongy condition, or may even not adhere at all if the substrate is unsuitable or if the current density is too high. Also, solution of the deposited material may occur. Such factors as these need prior investigation before the electrolytic method is relied upon. The difficulties are well brought out in an early paper by C. STATESCU[6].

(c) *Resistance measurement*

In general, the uncertainties surrounding the relation between electrical characteristics and thickness for thin metal or semi-conductor films are so great as to make this method quite useless. The resistivity of metals in thin layers is generally found to be considerably higher than that of the bulk material. It depends markedly on the structure of the film—degree of crystallinity, etc.—which in turn depends, sometimes critically, on the conditions of preparation of the film and on the after-treatment. Thus for germanium films 300–400 Å thick, deposited on quartz by evaporation, the resistivity was found to decrease by a factor of 600 on heating for an hour at 525° C in vacuo. Chromium films of 100–200 Å thick deposited on a cold substrate are found to possess resistivities differing from the bulk figure by factors of 10^3 to 10^4. The electrical properties of such thin layers are generally so sensitive to the conditions of preparation that the method of first determining the resistivity vs. thickness relation, for the subsequent evaluation of thickness from resistance measurements, is still unlikely to be reliable to within a factor of two.

There is one field, however, in which resistance measurements may be successfully applied. In studying the growth of oxide films on thin layers of metals, the decrease in the thickness of the metal consequent on the formation of the oxide (for which the conductance is effectively zero) may be followed from the resistance change. This method is not subject to the difficulties mentioned above and requires only that the resistivity of the metal layer shall be uniform throughout its volume, even though it may differ from that of the bulk metal or even from sample to sample. Direct comparison between thicknesses determined by this method with those obtained interferometrically for oxide films on copper (references [7-9]) have confirmed the reliability of the resistance method for this type of measurement.

R. C. L. BOSWORTH[10] studied the electrical behaviour of electro-lytically-deposited nickel films. The conductivity vs. thickness graphs were parallel to, though displaced from, those for the bulk material for films of thickness > 0.5 micron. The film thickness was deduced from the mass of the film, obtained from the loss in weight of the cathode, on the assumption that the films have the bulk density and that the true surface area is equal to the apparent area. The results suggest that these assumptions are not entirely justified and may lead to errors of 10–20 per cent.

(d) *Other electrical methods*

We conclude this section on electrical methods with brief mention of two types of measurement which have been applied to thickness determination. For certain dielectric layers, largely oxide films on a metal substrate, the capacitance of the condenser formed between the substrate and a mercury layer above the film may be measured. The thickness is evaluated using the formula for a parallel-plate condenser. A similar difficulty to the one affecting the resistance method is that the dielectric constant of the film is required and that this may not be the same as the value for the bulk material. Furthermore, the evaluation of the thickness involves a knowledge of the area of the condenser plates. As mentioned above, the true surface area may differ appreciably from the apparent area so that uncertainty from this cause further vitiates the results. These difficulties are discussed by W. BÄR[11] and A. ALEXANDROW and A. JOFFÉ[12]. The method has been successfully applied to films of aluminium oxide by A. GÜNTHERSCHULZE and H. BETZ[13].

For very sparse deposits, corresponding to less than one atomic layer, the change in work function of a surface has been shown (H. MAYER[14]) to give a sensitive indication of thickness. Thus for a tungsten surface covered by cæsium atoms, the work function falls from 4·1 volts for the pure tungsten surface to 2·6 volts when an amount of cæsium corresponding to about a quarter of an atomic layer is present.

It will be noticed that most of the electrical methods described are fairly old. The more recently-developed optical methods to be described in the remainder of this chapter enable film thickness and refractive index measurements to be made with high precision and without the uncertainties which are present in many of the methods described above. It is for this reason that the descriptions above of the non-optical methods have been made brief.

5.4. OPTICAL METHODS

When a beam of light is incident on a thin film, the character of the light reflected and transmitted by the film differs from that for the uncoated substrate. Measurement of certain of the quantities which characterize the reflected or transmitted beams yield values of film thickness, refractive index and extinction coefficient, some of the methods giving one or two, others all three quantities. Methods giving complete information are generally difficult and laborious.

Several methods will be described which give the film thickness only or the thickness and refractive index for transparent films. The method chosen for a particular measurement will often depend on attendant conditions which may exclude methods which would otherwise be suitable. Thus although the multiple-beam interference methods yield film thickness with high accuracy, they prove inconvenient if it is desired to study the change in film properties with age.

Certain of the early methods used in thickness determinations gave the optical thickness from which metrical thickness was obtained by assuming the refractive index of the film to be the same as that of the bulk material. Recent measurements (see Sections 5.6 and 5.7 below) have shown that this assumption may lead to results widely in error.

The optical methods to be described involve measurements of quantities such as the following:

Interference Fringe Displacements.

Colour, or wavelength, at which intensity maxima or minima occur.

Intensity of light reflected or transmitted.

State of polarization of reflected light.

It is seen from the equations developed in Chapter 4 that, given the thickness and optical constants of a film, or system of films, the reflectance, transmittance and phase-changes may be explicitly calculated. Explicit expressions for the thickness, refractive index and extinction coefficient in terms of directly measurable quantities have been derived (H. SCHOPPER[15]) and are given in Section 5.7 below. The expressions are extremely cumbersome and the value of the thickness so derived is, owing to the periodic nature of the solution of the thin-film reflectance equations, uncertain to the extent of an integral number of wavelengths. The method requires that measurements be made both of the amplitude and phase of the light reflected at both air and substrate sides of the film and also those for the transmitted beam. For certain types of problem—e.g. the study of films on opaque substrates—this method cannot be applied. Various methods have been used to overcome difficulties such as these. Trial-and-error methods in fitting experimentally derived curves of reflectance vs. wavelength to those computed from ranges of values of optical constants have been used. Approximations have been developed for the determination of film thicknesses which are small compared with the wavelength used or for dealing with absorbing films for which the transmittance is low. Graphical methods similar to those given in Chapter 4 have been devised.

The methods involving the measurement of fringe displacement, particularly those devised by O. WIENER[16], S. TOLANSKY [17], etc., are simple experimentally and are capable of high accuracy (see Section 5.5). They give the metrical thickness directly and do not require a knowledge of the optical constants of the film. They may be applied with equal facility to transparent and absorbing films. In the spectrophotometric methods, there is the usual difficulty of making accurate measurements of absolute intensities, frequently over a wide range. In the polarimetric methods, we deal with *ratios* of intensities rather than with absolute values. Such relative measurements can be made with much higher precision than can measurements of absolute values. Both polarimetric and spectrophotometric methods are capable of yielding complete information about the film. The thickness, refractive index and extinction coefficient may be obtained, albeit with some labour. The polarimetric method can also be used to detect non-homogeneity in the film (Section 6.9).

The methods so far described in this chapter have enabled the film thickness, or mass per unit area, to be determined. We shall therefore describe first among the optical methods those which yield values of film thickness independently of the optical constants of the film material. These methods (Section 5.5) which are applicable to any type of film on a smooth substrate are followed (Section 5.6) by optical methods suitable for transparent films on transparent substrates and finally (Section 5.7) we shall deal with methods suitable for the general case of an absorbing film on an absorbing substrate.

5.5. MULTIPLE-BEAM INTERFERENCE METHODS

These methods, developed to a remarkable degree by Tolansky[17], use the interference effects arising when two heavily-silvered surfaces are brought into close proximity. Of great value in studying surface topography generally, these interference techniques may be applied simply and directly to the determination of film thickness. The accuracy attainable is high.

The types of interference fringe used in this work fall under two headings:

(a) Fringes of constant thickness ('Fizeau Fringes').

(b) Fringes of equal chromatic order ('F.E.C.O.').

Either may be used for thickness measurement. The former type are slightly simpler experimentally; the latter are in certain circumstances capable of slightly greater accuracy.

(a) Fringes of constant thickness

When a wedge of small angle is formed between unsilvered glass plates and illuminated by monochromatic light, broad fringes are seen (Fizeau, 1862) arising from interference between the light beams reflected from the glass on the two sides of the air wedge.

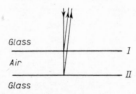

Figure 5.3   Interference in a film

At points along the wedge where the path difference between those two beams (allowing for phase changes at the surfaces) is an integral number of wavelengths, bright fringes are seen. Where the path difference is an odd number of half-wavelengths, dark fringes occur. Simple calculation shows that, for unsilvered surfaces, the amplitudes of the beams emerging after reflection at surfaces *I* and *II* (*Figure 5.3*) are nearly equal and that the amplitudes of beams which have suffered more than one reflection are negligible. The intensity distribution of light reflected therefore follows, approximately, a $\cos^2$ law as we move along the wedge. If part of one of the plates is covered by a transparent film, preferably with one edge lying parallel to the line of greatest slope of the wedge, the fringes are displaced on crossing the film edge and, from the fringe displacement, the film thickness may be found. The high vernier acuity of the eye enables very small displacements to be detected. The alignment of two sets of fringes with such a $\cos^2$ distribution may be effected to within a fortieth of a fringe. The measurement of a fringe displacement is subject to twice the error of setting a cross-wire on the centre of a single fringe and the over-all accuracy is about a tenth of a fringe. Furthermore, small variations in film thickness, occurring within a fringe width (which may extend over a millimetre or more) cannot be observed with this type of fringe.

106

If the surfaces of the plates are coated with highly-reflecting layers, one of which must however possess an observable transmission, then the reflected fringe system consists of very fine dark

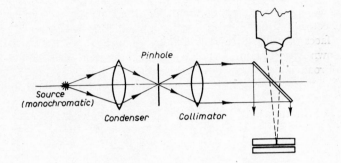

Figure 5.4   Apparatus for viewing Fizeau fringes

lines against a bright background.   The experimental arrangement is shown in *Figure 5.4.*   The fringe half-width may be made as small as 1/50 of the separation between orders so that fringe displacements may be made to this order of accuracy.   A typical fringe step, corresponding to a film of thickness 680 Å is shown in *Figure 5.5.*

The conditions for obtaining sharp Fizeau fringes have been examined by Tolansky[17] and J. BROSSEL[18].   The two most important factors in determining fringe sharpness are (1) the absorption of the reflecting layer and (2) the separation between the silvered surfaces.

The theory of the fringe shape for transmission fringes has been

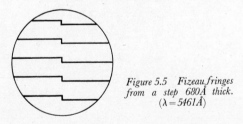

Figure 5.5   Fizeau fringes
from a step 680Å thick.
(λ = 5461Å)

given by Tolansky and for reflection fringes by J. HOLDEN[19].   The reflection case is complicated by the presence of an out-of-phase beam reflected at the glass/metal surface on the incident side. (Subsequent reflections occur at air/metal or film/metal surfaces and the phase changes on reflection differ from that at the glass/metal

107

interface.) For high values of reflection coefficient however, such as are used in high-resolution interferometry, the effect of this out-of-phase beam is negligible. The procedure in determining the intensity distribution in the fringe system is first to determine the dependence of intensity on the phase difference between successively reflected or transmitted beams for a parallel-sided film bounded by surfaces of reflectance R and transmittance T. This is the well-known Airy summation. The effect of a finite wedge angle is seen to broaden the fringes as given by the Airy summation. The conditions may be determined under which this broadening becomes negligible.

For transmission fringes, the Airy summation yields

$$I = \frac{I_{\max}}{1 + F \sin^2 \delta/2} \qquad \dots 5(3)$$

where $\delta = \frac{2\pi}{\lambda} . 2nt \cos \varphi$ is the phase difference between successively reflected beams in a film of refractive index n, thickness t and at an angle of incidence φ in the film. F is Fabry's 'coefficient de finesse', given by $\frac{4R}{(1 - R)^2}$ and $I_{\max} = \left(\frac{T}{T + A} \right)^2$ where A is the fraction of light absorbed in each metal film. Although the fringe *shape* is independent of the absorption, the contrast is seen to depend on A/T and forms the limit to the thickness of metal film which can be used.

The fringe half-width, expressed as the fraction of an order, is given by

$$W = \frac{1 - R}{\pi \sqrt{R}} \qquad \dots 5(4)$$

For vacuum-evaporated silver films, values of R up to 0·94 are attainable with which usable values of T are obtained. From 5(4) it is seen that fringe half-widths of only 1/50 of an order are realizable.

When interference takes place in a wedge, the multiple reflections do not all occur at the same value of film thickness t. Brossel has shown that the path difference between the first and r^{th} beams is given by $2rt - \frac{4}{3}r^3\theta^2t$ (where θ is the wedge angle). The rth beam will thus oppose the Airy summation if $\frac{4}{3} r^3\theta^2t = \lambda/2$. For the silver films mentioned above, the number of reflections effectively contributing to the interference effect is about 60 so that the largest permissible value of t may be found such that the intensity distribution is effectively given by the Airy summation. We find that for mercury green light, $t \not\gg 1/8X^2$ where X = number of fringes per cm of wedge. Thus with ~10 fringes per cm the film thickness should not exceed ~0.01 mm.

When the effects of errors of collimation and line width of source are investigated, it is found that large tolerances are again possible if the film thickness is kept to a small value. This enables large source 'pinholes' and high-pressure light sources to be used and so in turn thick, highly-reflecting silver layers to be employed.

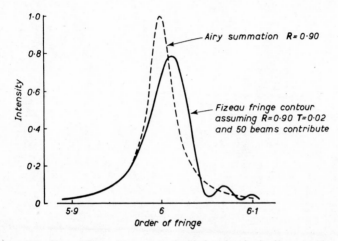

Figure 5.6   Form of Fizeau fringe compared with that given by the Airy summation

A more exact expression for the form of the Fizeau fringe contour has been determined by H. KINOSITA [20] who finds, for a sixth order fringe between surfaces of reflection coefficient 0.90 and transmission 0·02 that the maximum fringe intensity is 80 per cent of that given by the Airy summation and that this maximum occurs at an order 6·0125, instead of at exactly 6. Small subsidiary maxima occur on the high order side (*Figure 5.6*).

The sharpness of Fizeau fringe systems, especially that of the reflected system, may be considerably improved by the use of multiple dielectric films as reflecting layers (see Chapter 7). Successive quarter-wave pairs, alternately of high and low index, have a very high reflectance for the wavelength for which their optical thickness is $\lambda/4$ and possess very small absorption. It is the absorption in the reflecting film which sets a limit to the sharpness attainable in multiple-beam fringe systems.

By consideration of the Uncertainty Principle, E. INGELSTAM [21] shows that the accuracy of measurement on a single fringe step is limited to $\sim\lambda/1000$.

(b) *Accuracy of film thickness measurement using Fizeau fringes*

Multiple-beam Fizeau fringes are conveniently viewed with a low-power ($\times 50$) microscope. If higher powers are used, then a larger wedge angle is needed, so that a reasonable number of fringes

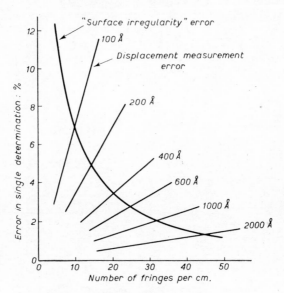

Figure 5.7   Curves enabling the optimum number of fringes per centimetre for typical optical surfaces to be obtained

are seen in the field of view; the fringes are then broadened owing to the violation of the phase condition mentioned in (a) above. The accuracy of locating the centres of fringes formed between silvered plates for which $R = 0.935$ and $F = 890$ and hence $W = 0.021$ of an order has been studied by O. S. HEAVENS [22]. For different fringe widths (obtained by varying the wedge angle) the standard deviation σ of a single setting was found to be related to the fringe width W_a by the relation

$$\sigma = 0.23\sqrt{(W_a)} \qquad \qquad \text{.... 5(5)}$$

both σ and W_a being measured in microns. Measurements were made on photographs of the fringe systems in order to avoid the danger of thermal drifts during measurements. This leads to an accuracy of better than 5 Å for a film thickness determined from a single step measurement, with a correspondingly smaller figure for the mean of several steps. This order of precision was found not to be realizable in practice owing to the local departure from planeness of optically-polished surfaces. Residual angles in the surfaces supporting the films cause variations in the fringe displacement at different parts of the surface and hence errors in the thickness determinations. The errors due to this cause are reduced by increasing the wedge angle; this, however, reduces the fringe displacement to be measured. An optimum wedge angle (or number of fringes per cm) exists such that the uncertainties due to residual surface angles and those due to fringe displacement measurements balance. From the intersection of the curves (*Figure 5.7*) giving the errors from these two sources as a function of the fringe separation, the optimum number of fringes per cm may be obtained. These considerations apply to films deposited on optical flats worked to $\lambda/5$.

(c) *Fringes of equal chromatic order*

Thickness measurements using fringes of this type (although using unsilvered surfaces) were first made by Wiener[16]. Instead of using a wedge, a parallel-sided film is used and illuminated with white light, which latter is subsequently dispersed. The resulting spectrum is banded and, if one of the plates is partly covered by a film, a displacement is seen in the fringe system. If the surfaces are unsilvered then an approximately $\cos^2$ intensity distribution with wavelength is obtained, with the limitations mentioned above. With heavily-silvered plates, the fringes become very fine. The experimental arrangement is shown in *Figure 5.8*. If a reflection system is used, a bright background is seen in the spectrum, crossed by very fine dark fringes. In transmission, the complementary pattern is seen. The fringes occur at wavelengths for which t/λ is constant so that, if the thickness of the space between silvered surfaces changes, the fringe moves to a new wavelength. For thickness measurement, normal incidence is used and the medium between the surfaces is air; thus $2nl \doteqdot 2l = m\lambda_1$ where m is the order of interference; this may be obtained from the wavelengths at which successive fringes occur. Thus if the fringe of order m lies at λ_1 and that of order $(m+1)$ at λ_2 we have

$$m\lambda_1 = (m+1)\lambda_2$$

so that
$$m = \frac{\lambda_2}{\lambda_1 - \lambda_2} \qquad \qquad \text{.... } 5(6)$$

The separation of the surfaces at that point is thus given by

$$2l = m\lambda_1 = \frac{\lambda_1\lambda_2}{\lambda_1 - \lambda_2} \qquad \qquad \cdots\ 5(7)$$

If the presence of a film of thickness t on one of the plates causes the fringe of order m to move to wavelength λ' then

$$\frac{2l}{\lambda_1} = \frac{2(l-t)}{\lambda'}$$

so that

$$t = \frac{(\lambda_1 - \lambda')l}{\lambda_1} \qquad \qquad \cdots\ 5(8)$$

This type of fringe possesses some noteworthy advantages over the Fizeau type fringe. Since the silvered surfaces are parallel, the Airy summation holds exactly and there is no fringe broadening from the wedge effect. Moreover, all the fringes in the spectrogram are produced by interference at one point of the film, this being the image on the film of the slit of the spectrograph, as produced by

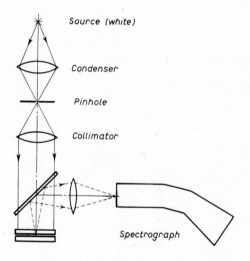

Figure 5.8   Apparatus for producing fringes of equal chromatic order

the projection lens in *Figure 5.8.* The latter lens may be a high-power microscope objective since the objections to the use of a high magnification mentioned in connection with Fizeau fringes do not apply to the equal chromatic order type. The results are not affected by residual surface angles as are Fizeau fringes and the source line-width is, of course, of no consequence. A carbon arc source enables heavy silverings to be used and hence high resolution to be obtained.

G. D. SCOTT, T. A. McLAUCHLAN and R. S. SENNETT [23] have applied this method to films of thickness down to 15 Å and have obtained an accuracy of ±5 Å, in agreement with the results of the last section when the effect of residual surface angles is omitted.

In using either of the types of multiple-beam fringe described above, reflection fringes must be used, the surface bearing the film being covered by an opaque layer of silver. This is to ensure that the reflection conditions above the film and above the unfilmed surface are the same. Use of too thin a silver layer may result in a difference in phase changes on reflection at these two regions. Tolansky [17] has shown that an opaque silver layer applied by vacuum evaporation contours accurately the surface on which it is deposited. Since the film is so covered, it is immaterial whether the film is transparent or absorbing.

5.6. MEASUREMENT OF THE THICKNESS AND REFRACTIVE INDEX OF TRANSPARENT LAYERS

A separate section is devoted to the case of transparent layers since there exist several simple methods for such measurements which are not applicable to the general case. Some of the methods to be given yield both refractive index and thickness whilst others give only one of these quantities or their product, the optical thickness. *Table 5.2* at the end of this section summarizes the range and scope of the methods described.

(a) *Early work on the colours of thin films*

As early as 1675 Newton realized that the colour of the light reflected by a thin film, illuminated by a parallel beam of white light, could be used to obtain a measure of the film thickness, if the refractive index of the film is known. With the notation of *Figure 5.3*, we see that the path difference between successively emerging reflected beams is given by $2n_1d_1 \cos \varphi_1 + \delta_0 + \delta_1$ where δ_0 and δ_1 are the path lengths corresponding to the phase changes on reflection at the upper and lower surfaces of the film respectively. If the film and substrate are both non-absorbing, then the phase changes are either 0 or π, giving path differences of 0 or $\lambda/2$. It is under this condition only that the method of estimating thickness from the colour of the reflected light can be simply applied. If either film or substrate are absorbing, then the phase changes are not zero or π but depend on the optical constants of the film and substrate.

If light is incident from a medium of index n_0 on to a film of index n_1 on a substrate n_2, then if n_1 lies between n_0 and n_2, the phase change on reflection at each surface of the film is equal to π and the path difference between successive beams is simply $2n_1d_1 \cos \varphi_1$. If this is equal to $m\lambda$, where m is an integer, then the reflected energy will show a maximum for that wavelength λ and the film will possess a characteristic hue.

If n_1 is greater than n_0 and n_2, there is a phase change of π at the n_0/n_1 surface and of zero at the n_1/n_2 surface. The reflected energy will thus show a minimum for a wavelength λ such that $2n_1d_1 \cos \varphi_1 = m\lambda$ and a maximum for a wavelength λ' such that $2n_1d_1 \cos \varphi_1 = (m - \frac{1}{2})\lambda'$.

Newton drew up a table 'wherein the thickness of films of Air, Water and Glass at which each Colour is most intense and specifick, is expressed in parts of an Inch divided into ten hundred thousand equal parts'. Interference up to the seventh order is considered and we note that, for this order, a water film $57\frac{3}{4}$ units thick appears Ruddy White.

The metrical thickness of the film is obtained, in terms of the refractive index, angle of incidence, etc., from the relation

$$d = \frac{m\lambda}{2(n_1^2 - \sin^2\varphi_0)^{\frac{1}{2}}} \qquad \dots \ 5(9)$$

where, if $n_0 < n_1 < n_2$, λ is the wavelength at which a maximum occurs in the reflected light. If n_1 is greater than both n_0 and n_2 then λ is the wavelength at which the minimum occurs. The order m may be found by first viewing the film at grazing incidence and, tilting so as to reduce the angle of incidence, counting the number of complete colour cycles up to a given angle of incidence. This will never be more than very few since, for films more than two or three wavelengths in (optical) thickness, the closing together of the interference orders results in a less well-defined hue. The change of phase of a light beam on reflection at a surface at non-normal incidence depends on the angle of incidence, the expression being given in equation 4(132). We notice, however, that for the wavelength at which successively reflected beams are in phase, or differ in phase by π, the phase change is always zero or π, whatever the state of polarization of the incident light. The maxima (or minima) for light polarized in or perpendicular to the plane of incidence thus occur at the same wavelengths for the two components. There is therefore no need to use polarized light when viewing the film. (This does not apply if the film is birefringent.)

For films of up to a few thousand ångströms optical thickness, this method is capable of yielding thickness values to within 10–15 per cent. A much higher accuracy may be obtained from the methods described in Sections (b)–(f) below, which enable the refractive index to be determined and do not depend on its being the same as that of the bulk material. A variation in the above method, in which the need for a subjective judgment of colours is avoided, is due to A. VAŠÍČEK [24]. Monochromatic light is used and the angle of incidence is varied, the values being recorded at which maxima or minima in the reflected beam occur.

When white light is used, the reflected light may be dispersed in a spectrometer and the wavelengths at which maxima and minima occur may be measured directly, instead of inferring them from the reflected hues. This method enables thicker films to be dealt with than the colour-judgment method since the order of interference may be obtained from the position of the bands in the resulting spectrum. Thick films will show many maxima in the reflected spectrum and the reflected light will appear white. So long, however, as the bands in the spectrum are not too close together to be resolved, the film thickness is easily obtained. Thus for the case of normal incidence (for which a reflecting plate at $45°$ between film and spectrometer is used) and for a film such that $n_0 \gtrless n_1 \gtrless n_2$, if the mth order maximum occurs at wavelength λ_1 and the $(m+1)$th order at λ_2, we have

$$2n_1 d_1 = m\lambda_1 = (m+1)\lambda_2 \qquad \dots 5(10)$$

whence

$$n_1 d_1 = \frac{\lambda_1 \lambda_2}{\lambda_1 - \lambda_2} \qquad \dots 5(11)$$

If the film has an index n_1 which is greater than both n_0 and n_2 the same equations may be used except that λ_1 and λ_2 are the wavelengths at which successive *dark* bands occur. If the bands are close together, it is clearly an advantage to use bands separated by several orders, rather than successive bands. Then if λ_1 is the wavelength at which the mth band occurs and λ_2 that of the $(m+p)$th band, we have

$$n_1 d_1 = \frac{p\lambda_1 \lambda_2}{\lambda_1 - \lambda_2} \qquad \dots 5(12)$$

115

Films on absorbing substrates.—The method described above may not be used directly for the determination of film thicknesses when the film lies on an absorbing substrate for in this case there is a phase change at the film/substrate surface which is a gruesomely complicated function of the optical constants of the substrate and film. A. B. WINTERBOTTOM [25] has shown by polarimetric methods that for cuprous oxide films on copper, this phase change corresponds to an optical thickness of film of 220 Å for wavelengths in the middle of the visible spectrum.

The limitations of the above methods are: (i) That the *optical* thickness is the quantity obtained. The refractive index must be known if the metrical thickness is to be determined. For very thin films there is evidence that the refractive index differs considerably from that of the bulk material. (ii) The film must be thick enough to show interference colours. Films of thickness of only a fraction of a wavelength cannot therefore be dealt with. We pass therefore to more recent methods of determining refractive index and film thickness which do not suffer from these limitations. The above methods are thus relegated to cases in which an approximate value of the optical thickness of a film is all that is required. For such purposes, these methods are, however, highly satisfactory and possess commendable simplicity.

(b) *Determination of n and d from measured reflectances at different wavelengths.*

F. ABELÈS [26] has given expressions enabling the two unknowns n and d of a film to be obtained from measurements of the reflectance of the film at two different wavelengths. Explicit expressions for n and d in terms of the measured reflectances cannot be obtained and the equations have to be solved by numerical methods of successive approximation. The calculations are not very elaborate when the light is polarized with the electric vector in the plane of incidence.

For normal incidence, if R_a, R_b are the reflectances for wavelengths λ_a, λ_b and if

$$A_a = \frac{1 + R_a}{1 - R_a} \qquad\qquad A_b = \frac{1 + R_b}{1 - R_b}$$

and
$$\alpha_a = \frac{4\pi}{\lambda_a} n_1 d_1 \qquad\qquad \alpha_b = \frac{4\pi}{\lambda_b} n_1 d_1$$

then with the help of equation 4(55) we obtain

116

$$(n_0^2 + n_2^2 - 2A_b n_0 n_2) \cos \alpha_a - (n_0^2 + n_2^2 - 2A_a n_0 n_2) \cos \alpha_b$$
$$= 2n_0 n_2 (A_a - A_b) \quad \ldots 5(13)$$

Since $\alpha_b = \dfrac{\lambda_a \alpha_a}{\lambda_b}$, this single equation for α_a may be solved (by trial-and-error). The value of n_1 may then be obtained from the quadratic in n_1^2

$$n_1^4 + \frac{(n_0^2 + n_2^2)(1 + \cos \alpha_a) - 4n_0 n_2 A_a}{1 - \cos \alpha_a} . n_1^2 + n_0^2 n_2^2 = 0 \quad \ldots 5(14)$$

d_1 is easily found from α_a and n_1.

The choice of wavelength at which the reflectances are measured is of considerable importance since for values of λ_a such that α_a is near to $m\pi$, the accuracy is low on account of the $(1 - \cos \alpha_a)$ term in the denominator of equation 5(14). Wavelengths corresponding to points about halfway down the reflectance curve on either side of the maximum are the most suitable. This suggests that the method should be used only if a monochromator is available. The film is conveniently deposited on a wedge-shaped ($\sim 3°$) plate of glass so that the light reflected from the back surface may be readily discouraged from getting in the way of the beam on which measurements are being made. A suitable experimental arrangement is described below.

For non-normal incidence, with the light polarized with the electric vector in the plane of incidence, equation 5(13) is replaced by

$$(n_0^2 + n_2^2 - 2n_0^2 \sin^2 \varphi_0 - 2n_0 n_2 A_b \cos \varphi_0 \cos \varphi_2) \cos \alpha_a$$
$$- (n_0^2 + n_2^2 - 2n_0^2 \sin^2 \varphi_0 - 2n_0 n_2 A_a \cos \varphi_0 \cos \varphi_2) \cos \alpha_b$$
$$= 2n_0 n_2 (A_a - A_b) \cos \varphi_0 \cos \varphi_2 \quad \ldots 5(15)$$

where, now

$$\alpha_a = \frac{4\pi}{\lambda_a} n_1 d_1 \cos \varphi_1 \qquad\qquad \alpha_b = \frac{4\pi}{\lambda_b} n_1 d_1 \cos \varphi_1$$

and φ_0, φ_1, φ_2 are the angles of incidence of the light in the media n_0, n_1, n_2 respectively. The equation corresponding to 5(14) becomes

$$n_1^4 + \frac{\{(n_0^2 + n_2^2)(1 + \cos \alpha_a) - 4n_0 n_2 A_a \cos \varphi_0 \cos \varphi_2 - 4n_0^2 \sin^2 \varphi_0\} n_1^2}{(1 - \cos \alpha_a)}$$
$$+ n_0^4 \sin^4 \varphi_0 + n_0^2 n_2^2 \cos^2 \varphi_0 \cos^2 \varphi_2 = 0 \quad \ldots 5(16)$$

117

For the case of the electric vector lying perpendicular to the plane of incidence, it becomes necessary to solve a quartic equation to obtain the value of n_1. This plane of polarization is therefore avoided when this method is used.

For the measurement of reflectances at normal incidence, the method described by P. ROUARD[27] and illustrated in *Figure 5.9* may be conveniently used. The film is deposited on a wedge-shaped plate P. Light from a monochromator is made parallel by the lens

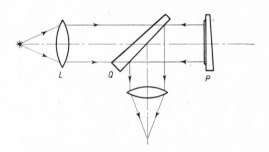

Figure 5.9   Arrangement of apparatus for measurement of reflectance.   (Rouard)

L and, after reflection by the film, is reflected by the similar wedge-shaped plate Q on to a suitable measuring device. Rouard used a photographic plate. A higher accuracy could now be attained by using a photocell in the regions of the spectrum in which such devices are sensitive. The detector is mounted on a spectrometer table so that the intensity of the direct beam may be measured. Care must be taken to ensure that the same part of the photocell surface is illuminated in each position of the detector. The use of a diffuser over the cell window minimizes errors from this source. The wedge plates enable the images formed by reflection at the back surfaces of P and Q to be excluded.

The experiment is not generally easy since the reflectances to be measured are usually of the order of a few per cent only. Also the thickness of the film needs to be known approximately since equation 5(14) possesses an infinity of solutions.

The corresponding treatment for an absorbing film on a transparent substrate is given in Section 5.7 below. Measurements on absorbing films must needs include the evaluation of phase changes on reflection at the surfaces and on transmission through the film.

(c) *Direct measurement of the refractive index of a transparent film by observing reflectance at the Brewster angle*

This method, due to Abelès[26], makes use of the fact that for light polarized with the electric vector in the plane of incidence the reflectance of a film of index n_1 on a substrate of index n_2 at the angle defined by $\tan \varphi_0 = n_1/n_0$ is the same as that of the uncovered substrate. This is easily seen to be the case from equation 4(44). Giving r_1, r_2 the values quoted in equation 4(29) and imposing the condition $\tan \varphi_0 = n_1/n_0$, the value of R reduces to $\tan^2 (\varphi_2 - \varphi_0)/\tan^2 (\varphi_2 + \varphi_0)$, which is precisely the value for the substrate alone. The method therefore entails adjustment of the angle of incidence until the intensity reflected by the film is the same as that of the substrate. Provided the index of the film does not differ too widely from that of the substrate, (± 0.3), an accuracy of ± 0.002 in the value of n_1 is attainable. This is probably the simplest of all methods for measuring the refractive index of a film. No calculation is involved and the measurement is independent of a knowledge of the refractive index of the substrate and of the thickness of the film. It is therefore of no consequence if the substrate is itself covered by a film or is optically inhomogeneous at the surface. No absolute intensity measurements are required if the refractive index only is required. The film must, however, be isotropic and homogeneous.

The thickness of the film may be determined from an additional measurement of the absolute reflectance of the film. From equation 4(49) we obtain

$$\cos 2\delta_1 = \frac{\cos 4\pi n_1 d_1}{\lambda} = \frac{r_1^2 + r_2^2 - R(1 + r_1^2 r_2^2)}{2 r_1 r_2 (1 - R)} \quad \dots 5(17)$$

where r_1, r_2 are the Fresnel coefficients and R is the reflectance at wavelength λ. d_1 is the only unknown. For the thickness determination the substrate index must be known.

(d) *Method using the Michelson interferometer (Fochs)*

The Michelson interferometer is used in a simple and elegant method devised by P. D. Fochs[28] which yields both film thickness and refractive index. The method is readily applicable to self-supporting films and to films supported on transparent substrates and has been used for cleaved mica sheets. The minimum thickness which can be dealt with is of the order of one or two wavelengths.

The experimental arrangement is shown in *Figure 5.10(a)*. White light from the Michelson interferometer, set with the mirrors M_1, M_2

accurately parallel, is directed on to the slit of a spectrograph, the slit and the mirrors being at conjugate foci of the projection lens. The film to be measured is arranged to cover two-thirds

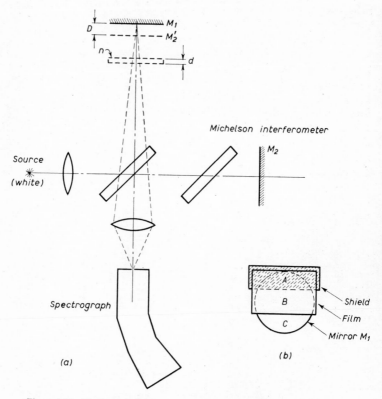

Figure 5.10 *Foch's method for measurement of film thickness and refractive index*

of the aperture of the mirror M_1, the film edge being set perpendicular to the direction of the spectrometer slit. Razor blades serve to cover one-third of the mirror apertures as shown in *Figure 5.10 (b)*. Three systems of fringes are seen in the field of view. In region A, where the interferometer mirrors are covered, a channelled spectrum is seen, arising from multiple reflections in the film. The relation between the order number m_1 of the dark bands, wavelength λ and the film thickness d and index n is simply

$$2nd = m_1\lambda \qquad\qquad \text{.... } 5(18)$$

120

In region C the normal Michelson fringes are seen, dark bands occurring at wavelengths given by

$$D = m_3\lambda \qquad \text{.... } 5(19)$$

In region B, the effect of the film is to change the path difference by $2(n-1)d$ so that the fringe pattern is given by

$$\{D + 2(n-1)d\} = m_2\lambda \qquad \text{.... } 5(20)$$

The value of D is then adjusted until the fringes in regions A and B are complementary, under which condition $D = 2d$. The thickness of the film is then found from equation 5(19), the order m_3 being determined in the usual way (see (a) above). The index is then determined from equation 5(18). In the event of appreciable dispersion, the two fringe systems cannot be brought into register over the whole spectrum. They are therefore aligned at that part of the spectrum for which the index is required. The accuracy of thickness measurement for films of thickness of the order 10 μ is ± 0.2 per cent; a precision of ± 0.003 is attainable in the refractive index determination.

(e) *Interference method for films of varying thickness (R. W. Stewart[29])*

The method described in (d) above requires that a reasonable area of parallel-sided film is available and employs equal chromatic order fringes. In the method used by Stewart, monochromatic light is used and fringes of equal thickness are utilized. By silvering the surfaces between which interference is to take place greater clarity and improved accuracy results. The non-uniform film is deposited on a silvered surface (this may be opaque since the observations may be made by reflected light) and a semi-transparent silver layer is evaporated over the film. The film should be non-uniform enough

Figure 5.11 Mask suitable for producing film for refractive index determination by Stewart's method

to give several fringes when viewed by monochromatic light. If an evaporated film is being studied, it is convenient to use an extended source and a mask of the form shown in *Figure 5.11* so that an extended 'hillock' is formed. The fringes seen on illuminating at normal incidence occur at points of equal optical thickness.

A semi-silvered flat is then placed close to the film when fringes of equal thickness of air film between the film surface and the flat are seen superposed on the fringes in the film itself. The appearance is as shown in *Figure 5.12*. Consider a section of the film by the plane *AB*, perpendicular to the plates of the system. Since *A* and *B* lie on the same fringe in the air film and on the same fringe in the film under investigation, the two path differences in both these

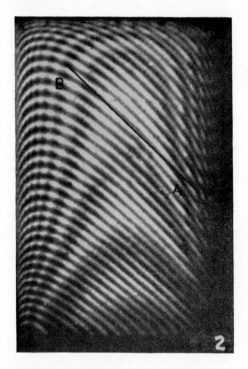

Figure 5.12   Appearance of fringes in Stewart's method

regions are equal. The section of the system by the plane through *AB* will therefore be as indicated in *Figure 5.13*, in which the vertical scale is enormously exaggerated. If corresponding to a layer of thickness *d* there are m_1 fringes in the air space and m_2 fringes in the film, we have $2d = m_1\lambda$ and $2nd = m_2\lambda$ so that $n = m_2/m_1$. The index is thus directly obtained by measuring the numbers of fringes of each type which cross a fixed length of the line *AB*. Stewart has used the method successfully to identify as cristabolite a layer deposited on the walls of a discharge tube during a sputtering process. An accuracy of ~ 0.01 is obtained.

A slightly modified version of this measurement employs a film in the form of a linear wedge. With a wedge of angle α (*Figure 5.14*) fringes in the film will occur where $2nd = m\lambda$ and their separation l_1 is given by $l_1 = \lambda/2\alpha n$. Fringes formed between the upper surface of the film and an adjoining flat (arranged with its line of greatest slope parallel to that of the film) will have a separation $l_2 = \lambda/2\beta$ whilst fringes between the uncovered portion of the lower flat and the second surface will be spaced $l_3 = \lambda/2\gamma$ apart. Hence

$$\alpha = \frac{\lambda}{2}\left(\frac{1}{l_3} - \frac{1}{l_2}\right) \text{ and}$$

$$n = \frac{l_2 l_3}{l_1(l_2 - l_3)} \qquad \text{.... 5(21)}$$

The chief difficulty with this method is to obtain a linear wedge of sufficient uniformity. The thickness should change by several wavelengths in a distance of a few millimetres across the surface.

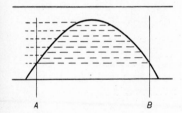

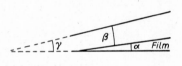

Figure 5.13   Section of film (Stewart)   *Figure 5.14   Wedge film method for n*

This is achieved by evaporating through a suitably shaped rotating sector. Fairly heavy silverings may be used, resulting in fine fringes so that the values of l_1, l_2 and l_3 may be measured to within $10\,\mu$ with ease.

(f) *Polarimetric methods.   Transparent films*

These methods are of application to both transparent and absorbing films and entail the examination of the state of polarization of light reflected by or transmitted by the film at non-normal incidence. In considering the reflection of plane polarized light from a surface we see from equations 4(19) and 4(21) that the amplitudes of the reflected components polarized in and perpendicular to the plane of incidence differ in magnitude and that there is in general a phase difference between them. The reflected light is thus elliptically polarized. By introducing a phase difference equal and opposite to that produced by reflection at the surface (Δ) plane polarized light is produced. The angle (ψ) between the vibration direction and the plane of incidence yields the ratio of the amplitudes of the reflected components in and perpendicular to the plane of incidence.

The presence of a film on the surface will in general change the values of Δ and ψ from those appropriate to the uncoated surface and, from the observed changes, the refractive index and film thickness may be found. Approximations have been derived by P. DRUDE [30] for the case $d \ll \lambda$. The method as applied by Vašíček is described below. For the general case explicit formulae for n and d cannot be obtained and methods of successive approximation are used.

When the film lies on a transparent substrate of a form which allows the state of polarization of the transmitted light, and that of the light reflected from the substrate side of the film, to be examined, then explicit formulae are obtainable, as shown by Schopper [15]. Since this method is of general application, suitable for dealing with absorbing as well as with transparent films, discussion will be deferred until Section 5.7 is reached.

Drude's approximations.—Drude derived approximate values for the change in the values of ψ and Δ, corresponding to the presence of a film on the surface, which are valid for values of d/λ small compared with unity.

The case considered is that of plane polarized light striking a surface, preferably only partly covered by a film, at an angle of incidence φ and with the plane of polarization inclined at an angle of 45° to the plane of incidence. The amplitudes of the components vibrating parallel and perpendicular to the plane of incidence are thus equal and the ratio of their amplitude reflection coefficients is given directly by the ellipticity of the reflected light. The expressions obtained by Drude, given below, require a knowledge of the optical constants of the underlying substrate. In view of the dependence of these values on the state of the substrate surface, it is necessary to determine the values of n_2 and k_2 for the surface used to support the film. If it can be arranged that the film covers only part of the substrate surface, then these measurements may be made at the same time as those of the film itself. The optical constants of the surface are calculated from the observed ψ, Δ values by well-established procedures. (See, e.g. [31-33] and also 5.7 (f)). If only a small area of surface is available, the method used by D. G. AVERY [34], is to be recommended.

124

The detailed derivation of Drude's approximation is lengthy and of doubtful fascination; the results only are given. If (ψ', Δ') are the values obtained for the uncoated surface and (ψ, Δ) those from the film, then

$$\psi - \psi' = \frac{2\pi}{\lambda} \frac{\sin 2\psi' \cos \varphi \sin^2 \varphi \, a_1}{(\cos^2 \varphi - a)^2 + a_1^2} (1 - n_1^2 \cos^2 \varphi)\left(1 - \frac{1}{n_1^2}\right)d_1 \quad \text{....} \quad 5(22)$$

$$\Delta - \Delta' = \frac{4\pi}{\lambda} \frac{\cos \varphi \sin^2 \varphi (\cos^2 \varphi - a)}{(\cos^2 \varphi - a)^2 + a_1^2}\left(1 - \frac{1}{n_1^2}\right)d_1 \quad \text{....} \quad 5(23)$$

where $\qquad a = \dfrac{n_2^2 - k_2^2}{(n_2^2 + k_2^2)^2}$ and $\quad a_1 = \dfrac{2n_2 k_2}{(n_2^2 + k_2^2)^2}$ $\qquad$ .... 5(24)

A. ROTHEN and M. HANSON [35] have examined the validity of the Drude equations for films of barium stearate on stainless steel and have shown close agreement for films of thickness up to 250 Å.

Vašíček's method [36].—The procedure employed by Vašíček is not restricted to the study of very thin layers. The same observations are made as in the Drude method and the thickness and refractive index are evaluated with the help of a graphical technique. The preliminary calculations are laborious but a high precision is attainable.

Let $R_p = \varrho_p e^{i\Delta_p}$ and $R_s = \varrho_s e^{i\Delta_s}$ be the amplitude reflection coefficients for the film-covered surface for the two planes of polarization. Then

$$\tan \psi = \varrho_p / \varrho_s \qquad \text{....} \quad 5(25)$$

and $\qquad\qquad\qquad \Delta = \Delta_p - \Delta_s \qquad\qquad \text{....} \quad 5(26)$

From equation 4(45) we have

$$R_p = \varrho_p e^{i\Delta_p} = \frac{r_{1p} + r_{2p}e^{-2i\delta_1}}{1 + r_{1p}r_{2p}e^{-2i\delta_1}} \qquad \text{....} \quad 5(27)$$

and

$$R_s = \varrho_s e^{i\Delta_s} = \frac{r_{1s} + r_{2s}e^{-2i\delta_1}}{1 + r_{1s}r_{2s}e^{-2i\delta_1}} \qquad \text{....} \quad 5(28)$$

where

$$\delta_1 = \frac{2\pi n_1 d_1 \cos \varphi_1}{\lambda}$$

Dividing 5(27) by 5(28) and separating real and imaginary parts, we obtain

$$\tan \Delta = \frac{AB' + A'B}{AA' - BB'} \qquad \dots \ 5(29)$$

where

$$\left.\begin{array}{l} A = r_{1p}(1 + r_{2p}^2) + r_{2p}(1 + r_{1p}^2) \cos 2\delta_1 \\ A' = r_{1s}(1 + r_{2s}^2) + r_{2s}(1 + r_{1s}^2) \cos 2\delta_1 \\ B = -r_{2p}(1 - r_{1p}^2) \sin 2\delta_1 \\ B' = r_{2s}(1 - r_{1s}^2) \sin 2\delta_1 \end{array}\right\} \qquad \dots \ 5(30)$$

For the azimuth,

$$\tan^2 \psi = \frac{(r_{1p}^2 + r_{2p}^2 + 2r_{1p}r_{2p} \cos 2\delta_1)(1 + r_{1s}^2 r_{2s}^2 + 2r_{1s}r_{2s} \cos 2\delta_1)}{(1 + r_{1p}^2 r_{2p}^2 + 2r_{1p}r_{2p} \cos 2\delta_1)(r_{1s}^2 + r_{2s}^2 + 2r_{1s}r_{2s} \cos 2\delta_1)} \qquad \dots \ 5(31)$$

These equations cannot be solved to give explicit expressions for n_1 and d_1 since all the r's involve n_1 and the $\cos 2\delta_1$ terms involve $n_1 d_1$. The procedure adopted is to choose likely values of n_1 and to use equation 5(31), which is a quadratic in $\cos 2\delta_1$, with the observed ψ to evaluate δ_1.

Equations 5(29) and 5(30) then enable Δ to be calculated for the chosen n_1. Repeating this procedure for a number of values of n_1 enables a curve of Δ vs. n_1 to be drawn from which the value of n_1 corresponding to the observed Δ may be obtained. A second curve of δ_1 against n_1 is drawn and the value of δ_1 corresponding to the true n_1 value, obtained from the Δ vs. n_1 curve, enables d_1 to be found. An idea of the precision which may be obtained is to be had from *Figure 5.15*, from Vašíček's paper.

The values of ψ and Δ obtained with films of various thickness and refractive index depend on the value of the substrate index so that a different set of computations is required for each substrate used. A. VAŠÍČEK [37] has given a table of values of ψ and Δ for films of thickness up to $1-1 \cdot 5 \ \mu$ and for indices ranging from $1 \cdot 2$ to $2 \cdot 75$ for a substrate of index $1 \cdot 5163$ (Schott glass no. BK$_7$) and for $\lambda = 5893$ Å.

Malleman and Suhner's method [38].—These workers use incident plane polarized light with the plane of polarization at 45° to the plane of incidence, as in Vašíček's experiments. They immerse the film in a liquid of index equal to that of the film itself under which condition the reflected light is itself plane polarized. The correct liquid is found by successive trials. Once the refractive index of the film is so determined, the film thickness is easily obtained, e.g. by measuring the ellipticity of the light reflected by the film in air.

The method involves the assumption that the properties of the film are not affected by immersion in the liquid; in view of the results of Schulz, described in section (g) below, this assumption appears to be justified, in so far as there is no evidence of a disturbance of the film structure. Owing, however, to the filling by the immersion liquid of the accessible porosities of the film, the index so

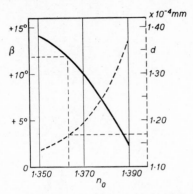

Figure 5.15 Derivation of n_1 and d by Vašíček's method

obtained will be the mean index of the film solid material plus that of the accessible voids when these are filled with the immersion liquid. The mean index of the same film in air will therefore be lower since the accessible voids are then filled with air. If the proportion of accessible voids is large, this effect may introduce considerable errors. Thus for LiF films of thickness 2μ, L. G. SCHULZ [39] finds that 30 per cent of the film consists of such voids, giving a mean index of 1·275. For this film thickness, there are no inaccessible voids so that the immersion method would yield the value of the index for the bulk LiF, viz. 1·392.

(g) The use of the Abbe refractometer

Before proceeding to the more sophisticated methods which are necessary for dealing with absorbing films and films on absorbing substrates, we may note that the standard Abbe refractometer has been used to yield results of high precision with films of thickness 1—2 μ and over. The work of Schulz [39] shows the remarkable potentialities of the method, from which considerable information about film structure has been obtained.

The film, deposited on a glass plate (e.g. microscope slide) is attached to the prism of the refractometer by means of two narrow strips of Scotch Tape (*Figure 5.16*) so that liquids of various refractive index may be used to unite optically the film to the prism. With a liquid of index higher than that of the film, two critical angles are observed. The one is the normal angle associated with the solid film and is independent of the refractive index of the uniting liquid.

Figure 5.16   Arrangement of film for refractive index measurement with the Abbe refractometer

The inherent roughness of evaporated films makes it possible for light within the liquid film to graze the prism surface so that a critical angle corresponding to the liquid film itself is also seen.

On the assumption of a continuous, solid film, the refractive index obtained for the film should be independent of that of the uniting liquid. If, however, the film contains porosities which may be filled by the liquid, then the average refractive index of the film will be found to increase with increasing refractive index of the liquid. This is found to be the case for LiF and NaCl. It is found, moreover, that films of lithium fluoride contain voids which are not accessible to the uniting liquid and a study of the variation of effective index of the film with filling liquids of different indices enables the fraction of accessible and inaccessible voids to be determined.

For, let f_s = fraction of optical path occupied by solid LiF and inaccessible voids. Then, if n_s is the refractive index of the combined phase solid + inaccessible void and n_v that of the liquid filling the accessible voids, the effective index of the system is given by

$$n_m = n_s f_s + n_v(1 - f_s) \qquad \text{.... } 5(32)$$

From a plot of n_m against n_v (*Figure 5.17*) we obtain f_s. From the intersection of the experimental lines A–F with the line $n_v = n_m$ ($= n_s$ from 5(32)), we obtain the value of n_s. For the thicker films, n_s is found to be significantly lower than the index for bulk LiF (1·3919) so the presence of inaccessible voids is indicated. Of practical interest is the value of the effective refractive index of the film in air. This is given by

$$n = 1·3919(1 - f) + 1·0002f \qquad \text{... } 5(33)$$

where f is the total fraction of porosities, both accessible and inaccessible. *Table 5.1* shows the results obtained for films of thicknesses ranging up to 180,000 Å. Two features of interest and importance are shown by these results. (1) The mean index of an evaporated film may be considerably lower than that of the bulk material. The usefulness of this property, e.g. anti-reflecting

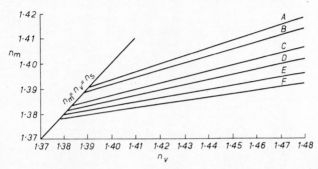

Figure 5.17 Experimental results of the measurement of refractive index of lithium fluoride films

films (see Chapter 7), is immediately realized. (2) The mean index does not attain the bulk value even for thicknesses as great as 18 μ, in contradistinction to the behaviour observed in many metal films, for which the optical constants are in fairly close agreement with those of the bulk metal for thicknesses of a few hundred ångströms.

The internal consistency of the results obtained by Schulz, and the way in which they agree with the general picture provided by the electron optical examination, suggest that the objection to immersing the film in a liquid, on the grounds of a possible disturbance to the film structure, is probably unfounded.

Table 5.1

Curve	Thickness in ångströms	n_m	Inaccessible voids in percent	Accessible voids in percent	Solid LiF in percent	Effective Index (n)
A	20,000	1·3915	0·0	30·0	70·0	1·275
B	35,000	1·3890	0·7	28·5	70·8	1·279
C	100,000	1·3842	1·3	24·2	74·5	1·295
D	135,000	1·3825	1·9	20·2	77·9	1·305
E	165,000	1·3795	2·7	16·4	80·9	1·317
F	180,000	1·3780	3·1	14·4	82·5	1·323

Solid LiF : $n = 1·3919$

(h) *Discussion and summary of methods for dealing with transparent films*

The most sensitive of the methods described above is the polarimetric one (Section 5.6(f)). A more complete discussion of this method is given in Section 5.7 below since the method is readily applicable to the determination of the optical constants of absorbing materials. It may also be applied (Section 6.9) to the investigation of inhomogeneity in films. Although the index and film thickness cannot be directly determined from the observations, the computations are not unduly laborious. As indicated below, considerable care in the experimental arrangements is required if the highest sensitivity is to be attained.

Among the simplest methods experimentally and one which yields an accuracy of ±0·002 in the refractive index is that due to Abelès (5.6(c)) involving the Brewster angle determination. For measurements on evaporated films it is doubtful whether a higher accuracy than this is needed on account of the dependence of the optical properties of such films on the conditions of preparation, ageing, etc. The Brewster angle method is superior to that which uses the measured reflectance at two wavelengths (5.6 (b)) since the latter requires the determination of reflectance absolutely. It must be stressed, however, that for the determination of film thickness an absolute reflectance measurement is necessary in conjunction with the Brewster angle method.

Of comparable accuracy in the index measurement, and of higher accuracy in the thickness determination is the Michelson interferometer method of Fochs[28]. It is suggested that if the refractive index alone is required, then the Abelès Brewster angle measurement is the most convenient; whereas if the film thickness is required as well as the refractive index, the method of Fochs is to be preferred if the film thickness is sufficiently great. If the films are too thin for this method to be practicable and if there is no objection to silvering the films then their refractive index could be found by the Brewster angle method and the thickness measured by means of multiple-beam interference. Under favourable conditions these methods yield results to better than ±10 Å.

The above methods require that there exist a reasonable area (≪a few square millimetres) of uniform film. The interference method described by Stewart[29] enables the refractive index to be found when the film thickness is non-uniform. The Schulz method using an Abbe refractometer may also be used for non-uniform films. In the last two cases the film thickness needs to be at least a few wavelengths.

130

Table 5.2 summarizes the relevant features of the various methods for the determination of film thickness and/or refractive index.

Table 5.2.   *Summary of Methods of Determining Refractive Index and Film Thickness*

Method	Suitable for	Range	Accuracy	Remarks
Weighing	t (mass/unit area)	1–2 atom layers	10^{-7} gm	
Molecular Ray	t (mass/unit area)	Fraction of atom layer		Possible uncertainty over condensation conditions.
Ionization	t(mass/unit area)	Fraction of atom layer		
Electrolysis	t (mass/unit area)	Fraction of atom layer		Electrolytes only. Adhesion and resolution problems.
Resistance	t			Unreliable for thin layers.
Reflected colour	nt	$\sim\lambda/4$ to few λ	$\pm 10\%$	Transparent substrate
Multiple-beam (Fizeau)	t	25 Å to few λ	± 10 Å	
Multiple-beam (F.E.C.O.)	t	15 Å to few λ	± 5 Å	
Reflectance at different λ's	n, t	few 100 Å to several wavelengths	variable	Entails absolute reflectance measurement
Michelson Interferometer	n, t	1–2 λ's to several wavelengths	$n : \pm 0.003$ $t : \pm 0.2\%$	
Brewster Angle	$n, (t)$	few 100 Å to several wavelengths	$n : \pm 0.002$ $t : 1\text{–}2\%$	$n_1 \sim n_2 \not> 0.3$. t requires absolute reflectance meast.
Polarimetric	n, t	<100 Å to several wavelengths	$n : \pm 0.0005$ $t : \pm 0.2\%$	Computation laborious
Abbe Refractometer	n	2 μ upwards	$n : \pm 0.001$	

5.7. MEASUREMENTS OF THE THICKNESS AND OPTICAL CONSTANTS OF ABSORBING LAYERS

Measurements of the optical constants of absorbing materials is somewhat difficult, as is illustrated by wide differences to be found in the values quoted by different observers.   In the case of bulk materials, these differences are probably related to the state of the surface of the specimen under examination, on which there may be an amorphous polish layer or an oxide or contaminating film.   Since for a strongly absorbing substance the depth of penetration of incident light is very small, considerable errors in measurements of this type may be introduced by minute amounts of disorder on the specimen surface.

Further difficulties arise when the material under investigation is in the form of a thin film. In early measurements of optical constants, for which it was assumed that the film behaved as a thin slice of bulk material, values obtained for the optical constants were found to differ widely from those of the bulk. It is now established that such apparent variations are due to the aggregated structure which is observed in matter in thin layers. The nature of the aggregation is such as to make exact correlation with the bulk

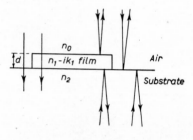

Figure 5.18

optical properties somewhat difficult. Progress has been made, however, and the results quoted below illustrate the essential correctness of the proposed theory.

Determination of the optical constants from the observations made is difficult, although in this respect investigations on thin films possess the advantage over those on bulk materials that the light transmitted by the film and that reflected by the film on the substrate side may be measured whereas for bulk specimens the reflectance at the air side is the only accessible quantity. If the film is deposited on a transparent substrate of known refractive index, then measurements of the complex amplitudes of the light reflected at normal incidence at both sides of the film and transmitted by the film enable the optical constants and thickness of the film to be evaluated explicitly. With films on absorbing substrates, exact solution is not possible and approximation methods are used. Graphical methods have been extensively used for these problems. For films which are so thick that multiple reflections may be neglected the standard techniques developed for measurements on bulk materials may be applied. (See R. Tousey [40] and D. G. Avery [34].)

(a) *Schopper's method*[15]

Measurements are made of the six quantities specifying the light beams transmitted by and reflected by both sides of the film. The theory is given for normal incidence. Although reflectance measurements are not readily possible at normal incidence, the errors introduced by the use of slightly non-normal incidence are unlikely to exceed the inherent uncertainties in the intensity and phase measurements. As with any method involving measurement at one wavelength the derived thickness value is uncertain as to the whole number of half-wavelengths.

The system is shown in *Figure 5.18*. With the notation in Chapter 4 and from the equations developed therein, we write down expressions for the transmittance and reflectances of the system. We denote by $S_0 e^{i\varphi_0}$ and $S_2 e^{i\varphi_2}$ the reflectances at the air (n_0) side and substrate (n_2) side respectively and by $De^{i\varphi}$ the transmittance of the system. From equations 4(45) and 4(46) we have

$$S_0 e^{i\varphi_0} = \frac{r_1 + r_2 e^{-2i\delta_1}}{1 + r_1 r_2 e^{-2i\delta_1}} \qquad \dots\ 5(34)$$

$$S_2 e^{i\varphi_2} = -\frac{(r_2 + r_1 e^{-2i\delta_1})}{1 + r_1 r_2 e^{-2i\delta_1}} \qquad \dots\ 5(35)$$

(the negative sign arising from the fact that, for light travelling in the reverse direction from that implied in equation 5(34), the Fresnel reflection coefficients are negative).

$$De^{i\varphi} = \frac{(1 + r_1)(1 + r_2)e^{-i\delta_1}}{1 + r_1 r_2 e^{-2i\delta_1}} \qquad \dots\ 5(36)$$

where
$$\delta_1 = \frac{2\pi(n_1 - ik_1)d_1}{\lambda}.$$

The quantities φ_0, φ_1, φ_2 representing the changes in phase on reflection and transmission are not readily measurable. The quantities more readily accessible are the differences in phase of the light reflected by the film and that reflected by the uncoated substrate and the corresponding quantities for the transmitted beams. Calling these quantities ε_0, ε_2 and ε and assuming that $n_0 < n_2$, in which case there is a phase change of π on reflection from the air side at the uncoated substrate, we readily see from *Figure 5.18* that the φ's and ε's are related by

$$\varepsilon_0 = \varphi_0 - \pi + \frac{4\pi n_0 d_1}{\lambda} \qquad \dots\ 5(37)$$

133

$$\varepsilon = \varphi + \frac{2\pi n_0 d_1}{\lambda} \qquad \qquad \text{.... } 5(38)$$

$$\varepsilon_2 = \varphi_2 \qquad \qquad \text{.... } 5(39)$$

Equations 5(34) to 5(36) may then be written

$$R_0 \equiv S_0 e^{i\varepsilon_0} = -\frac{e^{2i\gamma}(r_1 + r_2 e^{-2i\delta_1})}{1 + r_1 r_2 e^{-2i\delta_1}} \qquad \qquad \text{.... } 5(40)$$

$$R_2 \equiv S_2 e^{i\varepsilon_2} = -\frac{(r_2 + r_1 e^{-2i\delta_1})}{1 + r_1 r_2 e^{-2i\delta_1}} \qquad \qquad \text{.... } 5(41)$$

$$T \equiv D e^{i\varepsilon} = \frac{e^{i\gamma}(1 + r_1)(1 + r_2) e^{-i\delta_1}}{1 + r_1 r_2 e^{-2i\delta_1}} \qquad \qquad \text{.... } 5(42)$$

where $\qquad \gamma \equiv \dfrac{2\pi n_0 d_1}{\lambda}.$

The exponential factors containing γ and δ_1 are easily eliminated from equations 5(40) to 5(42). Writing the Fresnel coefficients r_1, r_2 in terms of the refractive indices, we obtain an expression for $\boldsymbol{n}_1$ involving the observable quantities.

$$\boldsymbol{n}_1^2 = n_2 \frac{n_2^2 S_0 e^{i\varepsilon_0}(1 + S_2 e^{i\varepsilon_2})^2 - n_0 D^2 e^{2i\varepsilon}[n_0 - n_2(1 + S_2 e^{i\varepsilon_2})]}{n_2 S_0 e^{i\varepsilon_0}(1 - S_2 e^{i\varepsilon_2})^2 + D^2 e^{2i\varepsilon}[n_2 - n_0(1 - S_2 e^{i\varepsilon_2})]} \qquad \text{.... } 5(43)$$

$$= A + iB$$

so that we obtain n_1, k_1 from

$$\left. \begin{aligned} 2n_1^2 &= (A^2 + B^2)^{\frac{1}{2}} + A \\ 2k_1^2 &= (A^2 + B^2)^{\frac{1}{2}} - A \end{aligned} \right\} \qquad \text{.... } 5(44)$$

Once the values of n_1, k_1 have been obtained, the film thickness may be readily calculated from equations 5(40)–5(42) which yield

$$\exp\left(\frac{4\pi i n_0 d_1}{\lambda}\right) = \frac{-S_0 e^{i\varepsilon_0} n_2 (\boldsymbol{n}_1^2 - n_0^2)}{\boldsymbol{n}_1^2 \{n_2 - n_0(1 - S_2 e^{i\varepsilon_2})\} + n_0 n_2 \{n_0 - n_2(1 + S_2 e^{i\varepsilon_2})\}} \quad 5(45)$$

Little is gained by giving the algebraic form of the expression for d_1, which is very cumbersome.

Schopper has used this method to study the optical behaviour of films of antimony sulphide, the thickness ranging up to about 2000 Å. The estimated accuracy of the figures for the refractive index and extinction coefficient is about ± 0.03 and for the thickness ± 30 Å. The values obtained for n (2·4–2·5 for the wavelength range 4360–6400 Å) are considerably below those reported for the bulk material. In dealing with the curves of reflectance and transmittance vs. thickness of Sb_2S_3 over the range stated, a close fit is obtained with curves computed from appropriately chosen values of n and k. One surprising feature of these results is that a good fit is obtained over the thickness range 400–2000 Å using the same values of n, k (for a given wavelength) over the whole range. This indicates that, in contrast to the optical behaviour of lithium fluoride, sodium chloride, etc., reported by Schulz, the optical constants of Sb_2S_3 vary but little with thickness over this range and that the structure of films of this substance in evaporated form is much less granular than is that of the alkali halides examined.

The experimental aspects of Schopper's method.—Considerable experimental care is required particularly in the measurement of the phase

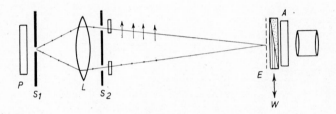

Figure 5.19 Arrangement of apparatus for phase measurement using Fleischmann and Schopper's method

of the reflected and transmitted beams. The amplitude measurement is made using the conventional photocell arrangement and is straightforward in principle. Since absolute values are required, however, close attention must be paid to this part of the measurement as well.

Of the several methods available for the phase measurement, the one described by R. FLEISCHMANN and H. SCHOPPER [41] appears to be the most satisfactory and yields measurements of the phase angle to within $\pm 0.4°$, leading to values of the optical constants to within ± 3 per cent.

The principle of the method may be easily followed from *Figure 5.19*. Light polarized by the polarizer P passes through a narrow slit and is converged on a quartz wedge by the lens L. A double slit S_2 is so arranged to produce a fringe system in the plane E in which the fringe separation is equal to that observed in the quartz wedge compensator when this is viewed between crossed polarizers.

Each aperture in S_2 is covered by a half-wave plate so that the directions of the planes of polarization of the two beams may be varied. They are adjusted so that the planes of polarization are perpendicular. The differential phase change in these two beams on passing through the quartz wedge W may be made to compensate for the phase difference arising from the different path lengths. This is effected by making the compensator fringe separation equal to that produced by the slit system. A uniform intensity over the whole field of view is thereby obtained. Introduction of the un-known phase difference into one of the beams, followed by re-setting the compensator for uniform brightness allows the phase difference to be read directly from the compensator scale. Fleischmann and Schopper describe a simple arrangement enabling this method to be used in conjunction with the Jamin refractometer.

As emphasized above, explicit expressions for the optical constants and film thickness involve the amplitudes and the absolute changes in phase of the light reflected from, or transmitted by, the film. Much work has been done, however, in which either the intensities only (photometric methods) have been measured or the ratios of amplitudes of, and phase differences between, the components of the light reflected by the film at non-normal incidence (polarization methods). With methods of these types, it is generally necessary to use approximations to evaluate the optical constants. Graphical methods have been devised to lessen the considerable labour involved.

(b) *Photometric methods. Murmann's treatment*

There are two general methods in this class. The first, used by H. MURMANN [42], requires that the thickness of the film be determined independently. In the second method, developed by Malé, the thickness may be deduced from the optical measurements although, as would be expected, the procedure is appreciably more complicated. In each case, the reflectance at each side of the film and the transmittance of the film are measured.

From the value of the film thickness and that of the wavelength to be used, values of R (reflectance on the air side) and R' (reflectance at the substrate side) are calculated for a range of values of n_1 and k_1. These are readily obtained from Section 4.9 (i)–(ii). For the calculation of R', n_0 is replaced by n_2 and n_2 by n_0. A typical curve is shown in *Figure 5.20 (a)*. From this curve we may read off the values of n_1, for each value of k_1, which correspond to the observed values of R and R'. Plotting against n_1 the values of k_1 which give

136

the values of **R** and **R'** obtained with the film in question then yields curves of the form shown in *Figure 5.20* (*b*). The method thus yields two possible pairs of values for n_1, k_1. When these values are fairly far apart, it is in general possible to distinguish the correct ones by calculation of the transmittance of the film and comparing the value with that observed. If the values are close together, this procedure may not indicate the correct value unambiguously.

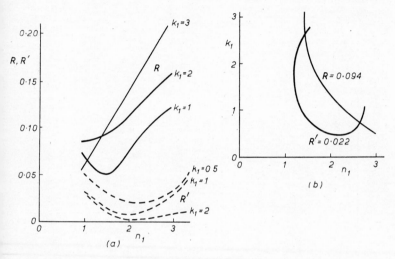

Figure 5.20   Typical curves obtained by Murmann's method for n and k

Errors due to the choice of the wrong values have in fact been reported. (See e.g. D. MALÉ [43].) For materials for which n_1 and k_1 are of the order 1–3, this method is satisfactory for a thickness range of 100–800 Å. For thicker films, the transmitted intensity becomes too small to measure accurately and the points of intersection of the curves of k_1 vs. n_1 become ill-defined. Smaller thicknesses are more easily dealt with using the method described after the next section.

In measurements of the optical constants of germanium films, W. H. BRATTAIN and H. B. BRIGGS [44] and H. A. GEBBIE [45] measured the transmittance of the films of several thicknesses and fitted the values of n, k by trial-and-error. This method is not unduly laborious once the values of n, k are reasonably close to the true ones and is reasonably sensitive, enabling n, k to be obtained to better than 3 per cent. Implicit in the method is the assumption that the optical constants of the film are independent of thickness. This may be true for an annealed film. For films condensed on a cold substrate, there is usually a dependence of the structure, and hence of the apparent values of n and k, on thickness.

(c) *Malé's method*[46]

As stated above, this method does not require an independent determination of the film thickness, but yields n_1, k_1 and d_1 in terms

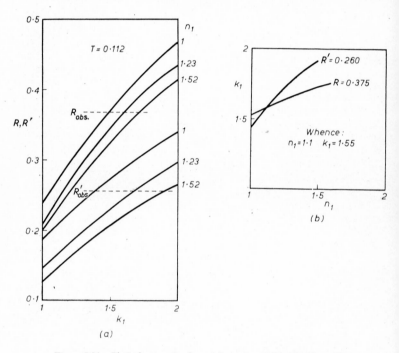

Figure 5.21   Typical curves for determining n, k and d by Malé's method

of the measured R, R' and T. The procedure involves heavy computation and is carried out as follows: A set of curves is prepared giving R, R' and T as functions of δ, where $\delta = \dfrac{4\pi n_1 d_1}{\lambda}$ for a (two-dimensional) range of values of n_1 and k_1. From each of these curves can be read off the calculated values of R, R' and δ, which occur at the observed value of T, for various values of n_1, k_1.

138

Curves are plotted of, e.g., R vs. k_1 for each value of n_1 used. (*Figure 5.21 (a)*.) From the latter curves, several pairs of values of n_1, k_1 may be read off corresponding to the observed value of R. These pairs of values are plotted (*Figure 5.21 (b)*). The whole procedure is repeated for R' and a second plot of k_1 vs. n_1 obtained corresponding to the observed value of R'. The intersection of the two curves yields the values of n_1, k_1 for the film.

With the values of n_1 and k_1 known, the value of d_1 is readily obtained from a plot of δ vs. n_1 for various values of k_1 from which, by interpolation, the value of δ corresponding to the values of n_1, k_1 found as above, is easily determined.

The labour involved in computing the curves of R, etc. vs. δ is heavy, although the computation (see Section 4.9, or 4.11 for graphical method) is straightforward. The same sets of curves are, of course, usable for any number of films so that their preparation is a 'once-for-all' task.

(d) *Approximate method for very thin films*

When the value of d_1 is known and is small compared with the wavelength used, the following approximations, due to H. WOLTER[47] and E. DAVID [48] may be used to evaluate the optical constants

$$n_1 k_1 = \frac{\lambda}{4\pi d_1} \cdot \frac{n_0 n_2}{n_2 - n_0} \frac{R - R'}{T} \qquad \dots\ 5(46)$$

together with

$$n_1^2 - k_1^2 = \frac{n_0^2 + n_2^2}{2} \pm \frac{\lambda}{2\pi d_1} \left\{ 2 n_0 n_2 \frac{R + R'}{T} - (n_2 - n_0)^2 \right.$$

$$\left. - \left(\frac{4\pi n_1 k_1 d_1}{\lambda} \right)^2 + \frac{n_1^2 k_1^2}{4} \left(\frac{4\pi d_1}{\lambda} \right)^4 (\mid n_1^2 - k_1^2 \mid + n_0 n_2) \right\}^{\frac{1}{2}} \dots\ 5(47)$$

The last two terms on the right of $5(47)$ are not known exactly since they involve n_1 and k_1; they are, however, small compared with other terms in the bracket. It suffices, therefore, to obtain an approximate value from equation $5(47)$ by ignoring the small last terms and to obtain the refined value of $n_1^2 - k_1^2$ by using this approximate value in the last terms of $5(47)$ as it stands.

(e) *Polarimetric methods. Absorbing films*

We deal in this section with the methods involving the measurement of the ratio of the amplitudes of the two components of plane polarized light after reflection at a film, and the differential phase change suffered by these components. As mentioned above, these measurements may be effected by observing the ellipticity of the reflected light and by measuring the azimuth. We do not, as in Schopper's method described above, deal with the absolute change in phase of the light on reflection. Explicit equations for n, k and d cannot be obtained in terms of the observables and curve-fitting or approximation methods are employed. These become more tractable when the film thickness is a small fraction of a wavelength. As early as 1890, Drude considered this problem and gave approximations valid for very thin layers. An extension to thicker films was made by Leberknight and Lustman and comparisons of these methods with exact treatment have been given by Winterbottom. In these methods the observations are made from the air side of the film only. Försterling and Essers-Rheindorf have treated the case of reflection from both air and substrate side, which may be used when the film lies on a transparent slab of substrate. The first-mentioned methods are useful for examining the surface films on opaque or absorbing substrates.

The problem of the accurate analysis of the state of polarization of light beams has received considerable attention in recent years. It is of general optical interest, rather than being of specific application to thin film problems. A detailed survey of methods would be out of place here, but some indication of the methods currently used may be of interest.

(f) *Methods for the analysis of elliptically polarized light*

We require to determine the ratio of the (real) amplitudes of the reflected plane-polarized components and the difference between their phase changes. With the notation of Chapter 4 (equation 4(37)), we have that $r_p = \varrho_p \exp i\Delta_p$ and $r_s = \varrho_s \exp i\Delta_s$ and we determine

$$\varrho = \varrho_p/\varrho_s \quad \text{and} \quad \Delta = \Delta_p - \Delta_s \qquad \text{.... } 5(48)$$

In most of the methods used in thin film investigations, plane polarized light with its vibration direction inclined at $45°$ to the plane of incidence is used. The reflected light is generally elliptically polarized and, since the amplitudes of the two incident components are equal, the azimuth is given by $\tan \psi = \varrho_p/\varrho_s$. If a phase difference equal and opposite to that occurring at the surface is introduced by a compensator, plane polarized light emerges from

140

the compensator and can be extinguished by an analyser. The analyser setting thus yields ψ and the compensator setting Δ.

The compensator may be one of fixed orientation and variable phase, e.g. a Soleil-Babinet type, or of fixed phase and variable orientation, such as the Senarmont Compensator. More elaborate methods, in which both fixed and moving compensators are used, have been described by W. VOIGT[49] and R. S. MINOR[50].

For the location of the plane of polarization of the light on leaving the compensator, electronic methods, using photo-electric cells as detectors, have been developed. Simple replacement of the eye by a photocell, in conjunction with an interrupted light source and tuned amplifier has been found to be effective (A. W. CROOK[51]). A rotating analyser and quarter-wave plate, suitable for measurements at the angle of principal incidence, has been used by C. V. KENT and J. LAWSON[52]. An ingenious photographic method, enabling measurements to be made simultaneously at several wavelengths, has been described by J. BOR[53]. Among the most recent photo-electric methods, with which azimuth and phase difference determinations may be made with an accuracy of one or two minutes of arc, is that given by J. F. ARCHARD, P. L. CLEGG and A. M. TAYLOR[54]. The methods used by the last-mentioned workers do not involve the use of a compensator, thus avoiding the need, with the conventional type of compensator, for separate calibrations at each wavelength used. An accuracy of the same order is obtained with the ellipsometer devised by A. ROTHEN[55] using a $\lambda/4$-plate with a half-shade system.

(g) *Methods for n, k using reflection from the air side of the film*

The amplitude reflection coefficient for light incident on the air side of a film ($\boldsymbol{n}_1$) of thickness d_1 on a substrate of index $\boldsymbol{n}_2$ (which may be complex) is given (equation 4(45)) by

$$R_{0(p,s)} = \frac{r_1 + r_2 e^{-2i\delta_1}}{1 + r_1 r_2 e^{-2i\delta_1}} \qquad \text{.... } 5(49)$$

For light polarized with the electric vector in the plane of incidence

$$r_1 \equiv r_{1p} = \frac{n_0 \cos \varphi_1 - \boldsymbol{n}_1 \cos \varphi_0}{n_0 \cos \varphi_1 + \boldsymbol{n}_1 \cos \varphi_0} \qquad \text{.... } 5(50a)$$

$$r_2 \equiv r_{2p} = \frac{\boldsymbol{n}_1 \cos \varphi_2 - \boldsymbol{n}_2 \cos \varphi_1}{\boldsymbol{n}_1 \cos \varphi_2 + \boldsymbol{n}_2 \cos \varphi_1}$$

For light polarized with the electric vector perpendicular to the plane of incidence,

$$r_1 \equiv r_{1s} = \frac{n_0 \cos \varphi_0 - \boldsymbol{n}_1 \cos \varphi_1}{n_0 \cos \varphi_0 + \boldsymbol{n}_1 \cos \varphi_1} \qquad \dots 5(50b)$$

$$r_2 \equiv r_{2s} = \frac{\boldsymbol{n}_1 \cos \varphi_1 - \boldsymbol{n}_2 \cos \varphi_2}{\boldsymbol{n}_1 \cos \varphi_1 + \boldsymbol{n}_2 \cos \varphi_2}$$

and the quantities φ_1, φ_2 are, for an absorbing film on an absorbing substrate, complex.

(h) *Drude's treatment*[56]

For films of thickness d_1, such that $n_1 d_1 \ll \lambda$, the exponents $(\exp - 2i\delta_1)$ may be replaced by $1 - 2i\delta_1$. The equations for ψ, Δ may be simplified by writing

$$A_{(p,s)} = r_2(1 - r_1^2)/(r_1 + r_2)(1 + r_1 r_2) \qquad \dots 5(51)$$

where the appropriate values for r_1, r_2 are inserted from 5(50a) or 5(50b) depending on the direction of the plane of polarization. We denote the amplitude reflection coefficient for the uncoated surface by r_{13} where

$$r_{13(p,n)} = \frac{r_1 + r_2}{1 + r_1 r_2} \qquad \dots 5(52)$$

Using the approximation indicated, we obtain

$$\frac{R_{0p}}{R_{0s}} = \frac{r_{13p}}{r_{13s}} [1 - 2i\delta_1(A_p - A_s)] \qquad \dots 5(53)$$

If ψ, ψ_0 are the restored azimuths with and without the film and Δ, Δ_0 the corresponding phase differences, then

$$\frac{e^{i\Delta} \tan \psi}{e^{i\Delta_0} \tan \psi_0} \doteq 1 + 2i\delta_1(A_p - A_s) \qquad \dots 5(54)$$

Writing $2i\delta_1(A_p - A_s) \equiv (a + ib)\dfrac{d_1}{\lambda}$ and noting that, for the small values of d_1 for which the approximation is valid, $\psi \simeq \psi_0$ and $\Delta \simeq \Delta_0$, we obtain

$$\psi - \psi_0 = \frac{ad_1}{2\lambda} \sin 2\psi_0 \qquad \dots 5(55)$$

$$\Delta - \Delta_0 = \frac{bd_1}{\lambda}$$

Subsequent work in which exact values of these quantities have been computed indicate that, for optical thicknesses of the order of 100 Å for typical materials, the error involved in using these forms is about 5 per cent. H. HAUSCHILD[57] has developed these expressions to include the second-order term in the expansion of $\exp(-2i\delta_1)$. The results are somewhat unwieldy.

(i) *Leberknight and Lustman's treatment*[58]

For films of too large a thickness to be covered by the Drude method, but in which the absorption is sufficiently high for reflections within the film (after the first) to be neglected, these authors derived expressions which give good agreement for film thicknesses up to about a half-wavelength. The condition under which the approximation holds is $r_2 \exp(-2i\delta_1) \ll r_1$ and the expression for the reflected amplitude is

$$R_0 \doteqdot r_1 + (1 - r_1^2) r_2 e^{-2i\delta_1} \qquad \text{.... 5(56)}$$

This method was used to study the growth of oxide films on metals as a function of the time of oxidation. Readings of $\tan \psi$ and Δ

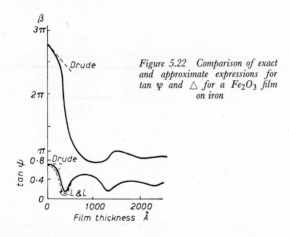

Figure 5.22 *Comparison of exact and approximate expressions for* $\tan \psi$ *and* $\triangle$ *for a* Fe_2O_3 *film on iron*

were taken at various stages of oxidation. The procedure for determining the film thickness was to determine theoretical curves of $\tan \psi$ vs. thickness and Δ vs. thickness using equation 5(49) (obtaining R_{0p} and R_{0s}, whence $\tan \psi$ and Δ are evaluated from their ratio) and to compare with the experimental curves found by

143

plotting $\tan \psi$, Δ against a linear function of the film thickness. In the work mentioned here, the square root of the time of oxidation was used. For work on evaporated or sputtered films, the time of evaporation would be used.

The ranges over which the above approximations are valid may be seen from *Figure 5.22*. The full lines show the values of $\tan \psi$ and Δ calculated from the exact expressions while the dotted lines give the approximation curves. The Drude expression is seen to be valid over a very small range only, particularly in respect of the value of $\tan \psi$. These curves were obtained for Fe_2O_3 films on iron by Winterbottom[25].

(j) *Försterling's treatment*[59]

For this method, measurements of the ellipticity of the light reflected at each side of the film and transmitted by the film are made. We thus obtain

$$e^{i\Delta} \tan \psi \qquad \text{(reflected at air side)}$$
$$e^{i\Delta'} \tan \psi' \qquad \text{(reflected at substrate side)}$$
$$e^{i\Delta_t} \tan \psi_t \qquad \text{(transmitted)}.$$

For the determination of the optical constants, we first evaluate

$$A = \frac{1 - \dfrac{\tan \psi'}{\tan \psi} e^{i(\Delta' - \Delta)}}{1 + \dfrac{\tan \psi'}{\tan \psi} e^{i(\Delta' - \Delta)}} \qquad \dots 5(57a)$$

$$B = \frac{\tan \psi}{\tan \psi_t} e^{i(\Delta - \Delta_t)} \qquad \dots 5(57b)$$

whence, for an angle of incidence φ_0, we write

$$y = \frac{2n_2^2}{n_2^2 - n_0^2}\left[\frac{1}{n_0^2 + (n_2^2 - n_0^2) \cos^2 \varphi_0} - \frac{1}{n_1^2}\right] \qquad \dots 5(58)$$

$$Q = \left(\frac{n_2^2 - n_0^2}{n_2^2 + n_0^2}\right)\frac{2(n_0^2 \sin^2\varphi_0 + n_2^2 \cos^2\varphi_0)}{n_0 n_2} B(A - 1) \qquad \dots 5(59)$$

n_1 may then be found from y by solving the equation

$$y - \frac{1}{y} = 2i \sin z \qquad \dots 5(60)$$

144

where $y = e^{iz}$. The thickness d_1 is evaluated from

$$\cot \left(\frac{2\pi n_1 d_1 \cos \varphi_1}{\lambda} \right) = -iA \frac{n_1^2 - 2n_2^2 \cos^2 \varphi_0}{(n_2^2 - 1)\sqrt{(n_1^2 - n_2^2 \cos^2 \varphi_0)} \cos \varphi_0} \qquad 5(61)$$

Försterling applied this method to the study of evaporated films of metal (the metal is not stated) and found that the solution of 5(61) yielded a complex value for d_1. The most likely interpretation of this anomalous result would seem to be that the films so prepared were not homogeneous and isotropic. The theory used in evaluating the optical constants and thickness is based on the assumption that the film is a homogeneous parallel-sided continuum. Films produced by vacuum evaporation are known to consist of aggregates of crystallites. If the crystallites are completely randomly oriented, then a result such as that obtained by Försterling would be unlikely since the film may be expected to behave simply as one of low density. If, however, there exist preferred orientation of the crystallites—a state observed even for films deposited on amorphous substrates (see reference[22])—then such anomalous results may well be expected. The defects of the theoretical treatment would be expected to be more serious for measurements made at non-normal incidence with such films. No such anomalies have been reported by Schopper[15] for antimony sulphide films, in which measurements were made at normal, or near-normal incidence.

(k) *Essers-Rheindorf's treatment* [60]

This author has derived approximations for use with the above method, which are valid for thin layers, the range of validity being determined by that of the expansion of $\exp(-2i\delta_1)$ (where $\delta_1 = 2\pi n_1 d_1 \cos \varphi_1/\lambda$) up to terms of the second order. The approximation is thus valid up to $n_1 d_1/\lambda \sim 0\cdot 1 - 0\cdot 2$. No anomalous results were obtained for thicknesses up to about 1000 Å. The method was compared with that using Newton's Rings. The films under observation were deposited over one-half of a flat, as shown in *Figure 5.23*. Corrections were applied to allow for multiple-reflections in the test film and the results of the two methods were shown to agree to within about 5 per cent. Fringe displacement measurements of this type are more easily and accurately made by the multiple-beam interference method, in which the use of an opaque silver layer removes the difficulties of multiple reflections in the test film and the uncertainties of phase differences at the two different surfaces at which the fringes are formed. (See Section 5.5).

(l) Avery's method

An interesting method for the determination of the ellipticity of the light reflected from a surface, and which has been applied to the determination of the optical constants of evaporated layers of silver, copper and tin, is given by D. G. AVERY [61]. Use is made of multiple-beam interference at non-normal incidence. Either fringes of equal thickness or fringes of equal chromatic order (Section 5.5) may be used. In the former type, maxima are observed in the transmitted

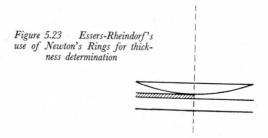

Figure 5.23   Essers-Rheindorf's use of Newton's Rings for thickness determination

light at points for which the path difference between successively-reflected beams is equal to an integral number of wavelengths. This path difference includes the phase changes on reflection at the silvered surfaces. For non-normal incidence, the phase changes depend on the relation of the plane of polarization to the plane of incidence. Thus doubling of the fringes occurs on changing the angle of incidence from zero. The appearance of Fizeau fringes between silvered plates with a reflectance of 90 per cent is shown in *Figure 5.24*. From the fringe separation the relative phase changes, for the components polarized parallel and perpendicular to the plane of incidence, may be found.

The method is shown to be less sensitive than the most accurate conventional polarimetric techniques for phase difference measurement. The experimental arrangement is, however, simple and the accuracy ($\sim 0.7°$) is adequate for many investigations concerned with thin films in which, owing to the sensitivity to conditions of preparation, structure, etc., reproducibility of results is not high. The range of thickness of films which may be dealt with is limited to that which will give reasonably sharp multiple-beam fringes.

(m) Schulz's method

Separate experiments for the determination of n and k were used by L. G. SCHULZ [62] on films of silver, gold, copper and aluminium prepared by thermal evaporation. The extinction coefficient is determined from the change in phase on reflection at the surface of the film. The phase change is measured by noting the difference in the wavelengths of the transmission bands of two interference filters (see Section 7.5) one of which has two silver layers and the other one layer of silver and one layer of the metal under test. The

$\theta = 0°$

40°

50°

60°

70°

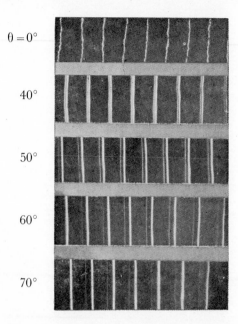

Figure 5.24   Appearance of multiple-beam interference fringes at non-normal incidence, used by Avery in the determination of differential phase changes

thickness of the dielectric separating the metal films is the same for each filter. Mica was used as the dielectric in Schulz's experiments. Phase change measurements were made with light polarized in and perpendicular to the plane of incidence and the extinction coefficient calculated from the differential phase change $\triangle$. The expression used for calculating the differential phase change is that corresponding to reflection at the bulk metal surface, viz.

$$\tan = \triangle \frac{2n_0 k_1}{n_1^2 + k_1^2 - n_0^2} \qquad \qquad .... \; 5(62)$$

where n_0 is the refractive index of the dielectric (mica) and n_1, k_1 the index and extinction coefficient of the metal film. For metals

147

in the visible and near infra-red regions, in which these measurements were made, n_1^2 is so small compared with k_1^2 that it may be neglected, or alternatively an approximate value suffices. Correction is made for the fact that the film is not quite opaque, amounting to some four or five degrees in $\triangle$ for films of thickness above 500 Å.

The index n is then obtained from measurements of the reflectance of a film-covered surface for each polarized component. Opaque films were used and several reflections at an angle of incidence of 45° enabled reflectance measurements, made with a Pulfrich photometer, to be obtained to within ± 1 per cent. The reflectances at 45° for the two polarized components are related by the equation $R_s^2 = R_p$ if reflection occurs under conditions which exactly satisfy the assumptions made in applying Maxwell's equations, namely that the film is homogeneous and that reflection takes place at a mathematically sharp, plane boundary. Freshly deposited films were found not to satisfy these conditions, the difference between R_s^2 and R_p being as much as 0·1–0·2. After suitable ageing and heat-treatment, the necessary homogeneity was attained. Values of n were obtained from films which satisfied the $R_s^2 = R_p$ condition by making use of the fact that for metals the large k-value enables the reflectance of the film to be written

$$R_0 = 1 - \frac{4n_0 n_1}{k_1^2} \qquad \qquad .... \; 5(63)$$

where R_0 is the reflectance at normal incidence. Furthermore, the value of R_0 lies fairly close to the value $\frac{1}{2}(R_s + R_p)$ for angles of incidence up to some 60°. The correction to be applied to the value of $\frac{1}{2}(R_s + R_p)$ to give R_0 may be easily calculated. From the reflectance measurements, together with the extinction coefficient determined by the method given above, the value of n_1 is readily found from equation 5(63). The value of n is sensitive to small errors in the reflectance, especially in cases where the extinction coefficient is large, as is apparent from equation 5(63). The method does not, however, require elaborate and expensive apparatus. For work with films, in which the results are somewhat structure-sensitive, the accuracy afforded by the method is generally sufficient.

5.8. The Choice of Method for the Measurement of Optical Constants

It will be clear to anyone who has contemplated the measurement of optical constants that this is no easy task. The apparent values of optical constants shown by a film or surface depend critically on the state of the first few layers. The measurements involved are not easy to make with high accuracy; and it would appear from much early work on the n, k values for thin films that different methods may yield results differing by unhealthily large factors from those of other people. Such large differences may, however, be generally attributed to differences in method of preparation of the film. Both sets of figures may well be correct for the different systems. If the film structures are widely different, agreement in n, k values would not be expected. Unless the state of the film is closely specified, the results for n, k are likely to be of little value unless determined by a method which takes account of the structure of the film.

From the known granular structure of thin films there is some uncertainty as to the real meaning of 'd', the film thickness which appears in the equations describing the film's optical behaviour. The uncertainty of density makes it undesirable that this quantity be determined by weighing. The methods to be preferred are clearly those in which the film thickness is determined simultaneously with the optical constants from the optical measurements. There are two main methods by which n, k and d may be found simultaneously. Schopper's method yields these values explicitly in terms of the measured amplitudes and phases of the reflected and transmitted beams. The calculation is fairly cumbersome but only one is required for a given set of results, in contrast to the method given by Malé in which sets of curves must first be drawn up giving reflectance and transmittance against thickness for a range of values of n and k. The computation in drawing up such curves is depressingly heavy. In contrast to the Schopper method, Malé's treatment requires only the amplitudes of the reflected and transmitted beams and these are readily obtained from the measured reflectances and transmittance. Schopper's method requires also the phases of the reflected and transmitted beams—a rather more difficult problem experimentally.

The complex thickness values obtained by Försterling may be due in some way to structural inhomogeneity in the film. No details are given of the conditions of preparation of the films on which measurements were made. Under certain conditions, fibre orientation occurs which may be accompanied by optical anisotropy. The effect of applying the equations derived for an isotropic film to such a film might well be to yield a complex value for the thickness. The methods in which normal incidence is used have not given such unrealistic results. For films prepared by evaporation at normal incidence, if a fibre axis develops it is itself normal to the substrate.

149

In measurements made at normal incidence, the light is travelling along the fibre axis direction. This is the one direction along which the effects of optical anisotropy in the film are of no consequence. Measurements made on such layers yield the propagation constants along the fibre axis.

For very thin films $(d/\lambda \ll 1)$ there is unlikely to be any trouble of the type mentioned above since such layers do not generally show any structural anisotropy. Non-normal incidence measurements, such as those of Drude, Leberknight and Lustman or Essers-Rheindorf, are likely therefore to be satisfactory.

For thicker films, Malé's method is highly suitable provided that a sufficiently extensive programme is contemplated to make worth while the preparation of the necessary curves. Otherwise, the Schopper method is likely to be the more rapid although the measurements are more difficult to make. The results obtained by the methods described are discussed in Chapter 6.

5.9. OTHER OPTICAL METHODS FOR DETERMINING FILM THICKNESS

Two other optical methods which have been applied to film thickness determinations may be mentioned. They are separated from the previous sections on account of the marked differences in technique which are used.

(a) *Tolansky's application of the light slit method*

The well-known Schmaltz light-slit technique has been applied to the measurement of film thickness for films of thickness above about one micron and yields an accuracy of about 2 per cent for films of thickness 10μ. The surface carrying the film (transparent or only slightly absorbing) is viewed under high magnification by oblique illumination and the image of a fine wire is superimposed on the image of the surface. (In the earlier applications of this technique, a fine slit covered the illuminating system, producing a fine illuminated line across the surface against a dark background.) Owing to the oblique illumination, two images of the wire shadow are seen, from reflection at the upper and lower surfaces of the film. From their displacement, and the known magnification of the system, the value of d/n, may be found where n is the refractive index of the film. An upper limit to the film thickness which can be dealt with is set by the fact that the two images cannot be focused simultaneously. The measurement necessitates the use of a high-power microscope objective with which the depth of focus is, inevitably, small. H. RAHBECK and M. OMAR [63] find that, even with slightly defocused images, consistent results to within 2 per cent are obtained with films of thickness 10 μ.

If the film index is not known, then an auxiliary experiment, in which the separation of fringes of equal chromatic order are observed, yields nd so that n and d are then both determined. This procedure is generally advisable on account of the uncertainty of the relation of the refractive index of materials in film form to that of the bulk material.

(b) *Friedman and Birks's X-ray method* [64]

Although this is in a certain sense a photometric method, the widely differing techniques involved in making measurements in this region of the spectrum provide reason enough for discussing this method separately from the visible-optical methods.

Use is made of the absorption of X-rays both in this and the following method. Friedman and Birks measure the intensity of one of the beams diffracted by the substrate before and after the deposition of the film. A Geiger-Müller counter is used for the intensity measurements. If a film of thickness d, mass absorption coefficient μ and density ϱ lies on a surface diffracting in a direction θ (Bragg angle), the path in the film (*Figure 5.25*) traversed by the diffracted beam is $d/\sin\theta$. If I_0, I are the intensities of the beams entering and emerging from the film, then

$$I = I_0 \exp\left(\frac{-2\mu\varrho d}{\sin\theta}\right) \qquad \text{.... } 5(64)$$

The mass absorption coefficient (μ) is readily obtained from a knowledge of the elements present. The density presents the usual thin film difficulty that the value may be less than that of the bulk material. The useful range of the method is, however, given as $10^{-5} - 10^{-2}$ cm. Only at the lower end of this range is there likely to be serious error from the density difficulty. *Figure 5.26* shows a suitable arrangement of apparatus, in which the principle of the X-ray focusing camera is employed.

(c) *Eisenstein's method* [65]

The measurements are in this case made on the lines diffracted by the film and those diffracted by the substrate. Theoretical treatments for the case of a film on a plane surface and also on a cylinder are given and the range covered is $10^{-6} - 5 \times 10^{-2}$ cm. At the lower end of the range the intensity ratio is of the order 1 : 100 so that the precision is not high.

For films containing heavy elements (from titanium upwards) use may be made of the fluorescent X-rays which can be excited by characteristic radiation of suitable wavelength. As the thickness

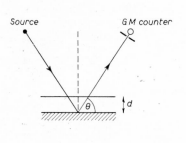

Figure 5.25

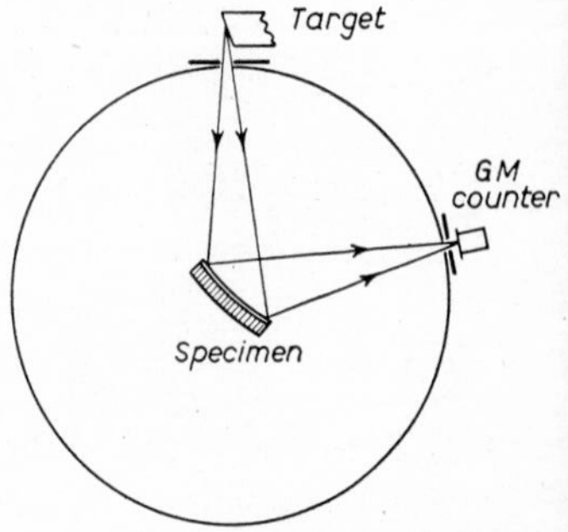

Figure 5.26   *Experimental arrangement in Friedman and Birks's method*

of a film increases, the intensity at the film surface of the fluorescent radiation from the deposited film increases whilst that from the substrate decreases. The ratio of the intensities thus forms a measure of film thickness. For chromium on iron, films of 500 Å upwards can be dealt with, the upper useful limit being a few microns.

5.10. USE OF RADIOACTIVE TRACERS

We close this chapter with the mention of a method which has become practicable since the ready availability of artificial radioactive materials. Since the high sensitivity of detection of radioactive elements, enables very small ($\ll 1\gamma$) masses to be measured, we have in principle a sensitive method for thickness determination. There are some difficulties, however, and the method is not absolute. The effect of self-absorption in the specimen film cannot be exactly allowed for, depending as it does on the film structure. Moreover, the method is applicable only to materials containing elements with suitably high capture cross-sections for the activating particle and with a suitably long half-period. The method has been applied to bismuth films by J. J. ANTAL and A. H. WEBER[66], calibration being effected by direct weighing on a microbalance. A marked dependence of the calibration on the surface roughness of the film was found, the rougher film giving the higher activity and giving roughly double the sensitivity compared with a smooth deposit. Thicknesses up to ∼300 Å could be dealt with accurately. Beyond this thickness the effects of self-absorption interfere decisively.

References

1 CLEGG, P. L. and CROOK, A. W. *J. Sci. Inst.* **29** (1952) 201
2 SENNETT, R. S. and SCOTT, G. D. *J.O.S.A.* **40** (1950) 203
3 BRADLEY, R. S. *J. Sci. Inst.* **30** (1953) 84
4 LANGMUIR, I. and KINGDON, K. H. *Proc. Roy. Soc.* **A 107** (1925) 61
5 BROCKMAN, F. G. *J.O.S.A.* **36** (1946) 32
6 STATESCU, C. *Ann. der Phys.* (4) **33** (1910) 1032
7 CONSTABLE, H. *Proc. Roy. Soc.* **A 115** (1927) 570
8 CONSTABLE, F. H. *Ibid.* **A 119** (1928) 196
9 DUNN, J. S. *Ibid.* **A 111** (1926) 203
10 BOSWORTH, R. C. L. *Trans. Far. Soc.* **30** (1934) 549
11 BÄR, W. *Z. Phys.* **115** (1940) 658
12 ALEXANDROW, A. and JOFFÉ, A. *Phys. Z. Sowj.* **2** (1932) 527
13 GÜNTHERSCHULZE, A. and BETZ, H. *Z. Phys.* **68** (1931) 145
14 MAYER, H. *Ann. der Phys.* (5) **33** (1938) 419
15 SCHOPPER, H. *Z. Phys.* **131** (1952) 215
16 WIENER, O. *Wied. Ann.* **31** (1887) 629
17 TOLANSKY, S. *Multiple Beam Interferometry*, O.U.P., 1948
18 BROSSEL, J. *Proc. Phys. Soc.* **59** (1947) 224, 235
19 HOLDEN, J. *Ibid.* **62** (1949) 405
20 KINOSITA, H. *J. Phys. Soc. Japan* **8** (1953) 219
21 INGELSTAM, E. *Ark för Fys.* **7** (1953) 309
22 HEAVENS, O. S. Ph.D. Thesis, Reading, 1951
23 SCOTT, G. D., McLAUCHLAN, T. A. and SENNETT, R. S. *J. Appl. Phys.* **21** (1950) 843
24 VAŠÍČEK, A. *Mitteilungen des Tschech. Akad. der Wiss.*, Prague, 1941
25 WINTERBOTTOM, A. B. *Trans. Far. Soc.* **42** (1946) 487
26 ABELÈS, F. *J. Phys. Rad.* **11** (1950) 310
27 ROUARD, P. *Propriétés Optiques des Lames Minces Solides* (Mémorial des Sciences Physiques, LIV)
28 FOCHS, P. D. *J.O.S.A.* **40** (1950) 623
29 STEWART, R. W. *Canad. J. Res.* **26** (1948) 230
30 DRUDE, P. *Wied. Ann.* **43** (1891) 126
31 TRONDSTAD, L. and FEACHEM, C. G. P. *Proc. Roy. Soc.* **145** (1934) 115
32 BOR, J. and CHAPMAN, B. G. *Nature* **163** (1949) 182
33 O'BRYAN, H. M. *J.O.S.A.* **26** (1936) 122
34 AVERY, D. G. *Proc. Phys. Soc.* **B 65** (1952) 425
35 ROTHEN, A. and HANSON, M. *Rev. Sci. Inst.* **19** (1948) 839; *Ibid.* **20** (1949) 66

36 VAŠÍČEK, A. *J.O.S.A.* **37** (1947) 145
37 — *Ibid.* 979
38 MALLEMAN, R. DE and SUHNER, F. *Rev. Opt.* **23** (1944) 193
39 SCHULZ, L. G. *J. Chem. Phys.* **17** (1949) 1153
40 TOUSEY, R. *J.O.S.A.* **29** (1939) 235
41 FLEISCHMANN, R. and SCHOPPER, H. *Z. Phys.* **131** (1952) 225
42 MURMANN, H. *Ibid.* **80** (1933) 161; *Ibid.* **101** (1936) 643
43 MALÉ, D. Thèse, Université, Paris, 1952
44 BRATTAIN, W. H. and BRIGGS, H. B. *Phys. Rev.* **75** (1949) 1705
45 GEBBIE, H. A. Ph.D. Thesis, Reading, 1952
46 MALÉ, D. *C. R. Acad. Sci., Paris,* **230** (1950) 1349
47 WOLTER, H. *Z. Phys.* **105** (1937) 269
48 DAVID, E. (see GOOS, F. *Z. Phys.* **106** (1937) 606)
49 VOIGT, W. *Phys. Zeit.* **2** (1910) 303
50 MINOR, R. S. *Ann. der Phys.* **10** (1904) 193
51 CROOK, A. W. Ph.D. Thesis, London, 1949
52 KENT, C. V. and LAWSON, J. *J.O.S.A.* **27** (1937) 117
53 BOR, J. *Proc. Phys. Soc.* **65** (1952) 753
54 ARCHARD, J. F., CLEGG, P. L. and TAYLOR, A. M. *Ibid.* 758
55 ROTHEN, A. *Rev. Sci. Inst.* **16** (1945) 26
56 DRUDE, P. *Wied. Ann.* **36** (1889) 532; *Ibid.* **39** (1890) 481
57 HAUSCHILD, H. *Ann. der Phys.* **63** (1920) 816
58 LEBERKNIGHT, C. E. and LUSTMAN, B. *J.O.S.A.* **29** (1934) 59
59 FÖRSTERLING, K. *Ann. der Phys.* **30** (1937) 745
60 ESSERS-RHEINDORF, G. *Ibid.* **28** (1937) 297
61 AVERY, D. G. *Phil. Mag.* **61** (1950) 1018
62 SCHULZ, L. G. *J.O.S.A.* **44** (1954) 357;—and TANGHERLINI, F. R. *Ibid.* 362.
63 RAHBECK, H. and OMAR, M. *Nature* **169** (1952) 1008
64 FRIEDMAN, H. and BIRKS, L. S. *Rev. Sci. Inst.* **17** (1946) 99
65 EISENSTEIN, A. *J. App. Phys.* **17** (1946) 874
66 ANTAL, J. J. and WEBER, A. H. *Rev. Sci. Inst.* **23** (1952) 424

6

RESULTS OF OPTICAL MEASUREMENTS
ON FILMS

Differences between film and bulk properties—The reflectance and trans-
mittance of single layers—The phase changes on reflection at and trans-
mission by single layers—The optical constants of thin films—Results
of measurements on gold films—Results of measurements on silver films
—Results of measurements on other metals—Absorption in thin films—
Optical inhomogeneity of films

6.1. DIFFERENCES BETWEEN FILM AND BULK PROPERTIES

MEASUREMENTS of the refractive indices of bulk transparent
materials present no serious difficulties. The effect on a transmitted
light beam of disturbance from any slight surface disorder, e.g. from
polishing, or from surface films, is generally negligible if the beam
traverses a total distance in the medium of the order of millimetres.
The state of polarization of light reflected by such a surface may well
be considerably affected by the state of the surface. For the
determination of refractive index, however, transmission methods
may be used and difficulties associated with surface films thereby
avoided.

The study of absorbing materials, and especially those in which
the absorption is large (e.g. metals in the visible and infra-red),
presents a much more difficult problem. For dealing with bulk
specimens of highly-absorbing materials, the only method open to
the investigator entails examination of light reflected from the specimen surface. Since the depth of penetration of the light-wave may
well be much less than a visible wavelength, it is apparent that the
behaviour exhibited by a specimen depends almost entirely on the
state of the surface which may be completely unrepresentative of
the state of the interior of the specimen. In much of the early work
on the reflecting power of metal surfaces, great pains were taken to
ensure a high degree of polish on the surfaces used. Grinding
followed by prolonged polishing resulted in a very smooth, even
surface but electron diffraction studies have revealed considerable
disorder of the normal crystal structure of the metal by such polishing
processes. With the growth of vacuum evaporation methods, there
appeared at first sight to be an excellent method for studying the

optical properties of materials in a pure state. Since evaporated deposits generally contour accurately the substrate on which they are deposited, one had simply to deposit the substance on an optically polished glass or quartz surface thereby obtaining the necessary smooth surface with no complication from the action of polishing the material itself. The optimism, that earlier difficulties had thereby been overcome, was short-lived. Materials in the form of evaporated layers proved to have very strong minds of their own and to display properties quite different from those of the bulk material. Far from escaping the difficulty of not having the surface in a defined state of order, one had the greater difficulty to face in which the state of the film, and in consequence its properties, depended markedly and in some cases very critically on the manifold conditions attending its preparation.

These points have been discussed at length in Chapter 2. We need to be reminded of them in the light of the often depressingly large differences in the values of optical constants obtained by different observers.

6.2. The Reflectance and Transmittance of Single Layers

(a) Transparent films

The expressions for reflectance and transmittance developed in Section 4.4 enable the curves of reflectance and transmittance of a film as a function of thickness to be calculated in terms of the refractive indices of layer and substrate. Typical curves, calculated for transparent films of various indices on a substrate of index $n_2 = 1 \cdot 5$ are shown in *Figure 6.1*. For a quarter wavelength film of lower index than that of the substrate, destructive interference occurs between the waves reflected from both sides of the film and a reflectance lower than that of the substrate results. For a film of index higher than that of the substrate, reinforcement occurs at an optical thickness of $\lambda/4$ and an enhanced reflectance results. Use is made of these effects in the preparation of anti-reflecting and high-reflecting films. (Chapter 7.)

The reflectance at normal incidence of a single film of thickness d_1 and refractive index n_1 on a substrate of index n_2 is given by equation 4(48)

$$R = \frac{r_1^2 + 2r_1r_2 \cos 2\delta_1 + r_2^2}{1 + 2r_1r_2 \cos 2\delta_1 + r_1^2 r_2^2} \qquad \text{.... } 6(1)$$

156

where $\qquad r_1 = \dfrac{n_0 - n_1}{n_0 + n_1}, \quad r_2 = \dfrac{n_1 - n_2}{n_1 + n_2} \quad$ and $\quad \delta_1 = \dfrac{2\pi}{\lambda} n_1 d_1.$

Replacing r_1 and r_2 respectively by $r_2' = \dfrac{n_2 - n_1}{n_2 + n_1}$ and $r_1' = \dfrac{n_1 - n_0}{n_1 + n_0}$

gives the reflectance of the system for light incident from the substrate side. Since $r_1' = -r_1$ and $r_2' = -r_2$, we see that the reflectance at normal incidence for light falling from either side on the film is

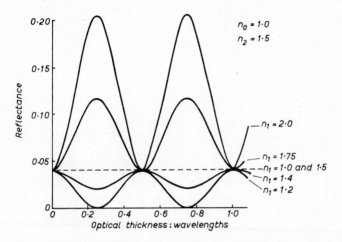

Figure 6.1 Variation of reflectance (at air side) with thickness for films of various refractive indices on a substrate of index 1·5

the same. Consideration of the case of non-normal incidence shows that this result is true for non-normal incidence. It is not, however, valid for an absorbing film.

From equation 6(1) it is seen that maxima and minima respectively of the reflectance curve occur at values of $n_1 d_1$ given by

$$n_1 d_1 = (2m+1)\lambda/4 \quad \text{and} \quad n_1 d_1 = (2m+2)\lambda/4$$

The values of the reflectance at these points are

$$n_2 \gtrless n_1 \gtrless n_0$$

$$(\boldsymbol{R})_{\min} = \left(\frac{r_1 - r_2}{1 - r_1 r_2}\right)^2 = \left(\frac{n_2^2 - n_0 n_1}{n_2^2 + n_0 n_1}\right)^2 \qquad \dots \ 6(2)$$

$$\text{for } n_1 d_1 = (2m+1)\lambda/4$$

157

$$(\boldsymbol{R})_{max} = \left(\frac{r_1 + r_2}{1 + r_1 r_2}\right)^2 = \left(\frac{n_2 - n_0}{n_2 + n_0}\right)^2 \qquad \qquad \dots 6(3)$$

$$\text{for } n_1 d_1 = (2m + 2)\lambda/4$$

$$n_2 \gtrless n_1 \lessgtr n_0$$

$$(\boldsymbol{R})_{min} = \left(\frac{n_2 - n_0}{n_2 + n_0}\right)^2 \quad \text{for } n_1 d_1 = (2m + 2)\lambda/4 \qquad \dots 6(4)$$

$$(\boldsymbol{R})_{max} = \left(\frac{n_2^2 - n_0 n_1}{n_2^2 + n_0 n_1}\right)^2 \quad \text{for } n_1 d_1 = (2m + 1)\lambda/4 \qquad \dots 6(5)$$

It will be noted that the maximum reflectance indicated by equation 6(3) (and the minimum by 6(4)) is independent of the refractive index of the film.

Referring to equation 4(44), giving the amplitude of the wave reflected at the film, we see that, under the maximal and minimal conditions given above, the reflected amplitude is real. If the value of $(r_1 + r_2 \cos 2\delta_1)/(1 + r_1 r_2 \cos 2\delta_1)$ is positive at the maximum or minimum, there is thus no change of phase in the reflected beam. If the ratio is negative, then a phase change of π is indicated.

Figure 6.2 shows the results obtained by K. HAMMER[1] for films of zinc sulphide for wavelengths in the visible spectrum. In this region, ZnS is practically non-absorbing, as is indicated by the almost constant values of maximum reflectance at different thicknesses. From the values of the thickness at which the maxima and minima occur, the refractive index of the layer may be readily found from equations 6(4)–6(5) above. The results are shown in *Table 6.1*.

Table 6.1. Refractive Index of ZnS

Wavelength (Å)	1st Max	1st Min	2nd Max	2nd Min	3rd Max
4046	2·41	2·46	2·42	2·44	2·44
4358	2·29	2·47	2·42	2·49	—
5460	2·28	2·29	2·34	—	—
7200	2·31	2·32	—	—	—

The discordance of these results, in which the spread is greater than would be expected from the accuracy with which the turning values can be located, is probably explained by the dependence of the film structure and, hence of its optical properties, on the thickness of the layer. It will be noted that the values of the index obtained from the thinner films are lower than those from thicker films. This general behaviour is in accord with the results of Schulz discussed in Section 5.6. The presence of voids, more marked in the thinnest films, gives rise to a mean index below that of the bulk material.

(b) *Absorbing films (absorption small)*

From the expressions developed in Chapter 4 for the reflectance (at each side of the film) and the transmittance of an absorbing film,

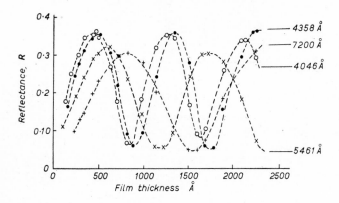

Figure 6.2 *Hammer's results for zinc sulphide at various wavelengths*

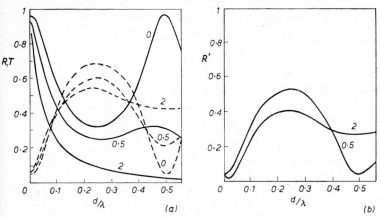

Figure 6.3 *Computed curves of* (a) **R**, **T** *and* (b) **R**¹ *vs. thickness for* $n_1 = 4$ *and various values of k*

curves may be obtained for the dependence of these quantities on film thickness. Typical curves are shown in *Figure 6.3*. The effect of absorption with increasing film thickness reduces the amplitude of successive maxima in the reflectance and transmittance curves and displaces them in the direction of smaller thicknesses compared with the transparent film. If the absorption is sufficiently high, as

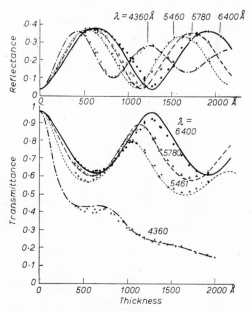

Figure 6.4 Schopper's results for antimony sulphide. The full lines are theoretical curves computed from the following values of n and k:

λ	n	k
6400	2·5	0·01
5780	2·5	0·04
5461	2·5	0·07
4360	2·6	0·29

is the case for metals in the visible region of the spectrum, the curve of transmittance vs. thickness falls rapidly and smoothly. With very high absorption, the amplitude of the beam within the film falls to an insignificant value even for an optical thickness of a quarter wavelength.

Figure 6.4 shows the reflectance (on the air side) and the transmittance curves for films of antimony sulphide deposited by thermal evaporation. Films of Sb_2S_3 are a deep orange in colour; their

160

increasing absorption with diminishing wavelength is shown by the curves of *Figure 6.4*. At $\lambda = 6400$ Å, the curves approximate to the form shown in *Figure 6.1*; at this wavelength the extinction coefficient is very small (<0.01). At $\lambda = 4358$ Å, the absorption is becoming heavy ($k \sim 0.29$) and curves similar to the theoretical form shown in *Figure 6.3* are obtained. The solid curves of *Figure 6.4* are calculated from the equations for reflectance and transmittance using values of n, k deduced from observations of the amplitude and phase of the light reflected (at each side) and transmitted by the film. (This method of obtaining optical constants is described in Section 5.7.) The slight discrepancy between theoretical and experimental curves probably indicates a variation of n and k with thickness.

(c) *Absorbing films. (Absorption large—metal films)*

The absorption of metals in the visible region of the spectrum is generally so high that the transmittance of a film a few hundred ångströms in thickness is of the order one per cent or less. The corresponding optical thicknesses at which appreciable transmission occurs are generally well below a quarter wavelength in this region so that interference maxima and minima are not observed. (The latter are to be obtained with semi-conducting materials in the near infra-red and even at the long wavelength end of the visible spectrum.)

The general form of the relation between the reflectances R, R' and the transmittance T and the thickness of a layer (assumed isotropic and homogeneous) is shown in *Figure 6.5*. R, R' and T are calculated for a platinum layer with optical constants n and k corresponding to the figures for bulk platinum. The substrate is assumed to be quartz ($n = 1.56$). The minimum in the R' vs. d curve, at one time thought to be an anomalous feature of thin films of metals, is seen to be precisely in accord with the theory of a thin homogeneous layer. Minima in the R' vs. d curve are observed for many metals. Some typical curves are given below. Although of the same general shape as the curves of *Figure 6.5* above, the experimental curves do not agree with those obtained by calculating R, R' and T using the bulk values of n and k. Nor, in fact, can a fit between theory and experiment be obtained by assuming any constant values of n and k. The effective optical constants of films of this order of thickness are found to be strongly dependent on the film thickness. This behaviour is further discussed in Section 6.4 below.

Platinum films

The dotted curve in *Figure 6.5* shows the experimental results obtained for sputtered platinum films by A. Partzsch and W. Hallwachs[2]. The difference from the theoretical curves calculated from the bulk constants is not so great as that obtained with many other metals although it is enough to show that the simple theory is

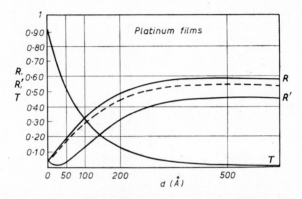

Figure 6.5   *Reflectance and transmittance curves for platinum calculated from the optical constants of the bulk metal.   Dotted curve: experimental results on sputtered films of platinum (Partzsch & Hallwachs[2]).*

inadequate. The smaller discrepancy for sputtered platinum suggests that the film structure for this metal is less granular than that obtaining in silver and gold films. An interesting feature of the **R'** vs. *d* curve for platinum is the insensitivity to wavelength of the position of the minimum of the curve. Thus multiple reflections within a glass plate may be suppressed, even for white light, by thin (25 Å) platinum films sputtered on each side of the plate. Similar behaviour in this respect is observed for thin layers of aluminium, beryllium and zinc. Films of gold, silver, copper and iron also show a minimum in the **R'** vs. *d* curve but the position of the minimum is sensitive to wavelength.

Gold films

There have been several investigations on the reflectance and transmittance of thin gold films, in the course of which both beaten gold (suitably thinned) and evaporated films have been examined. Large differences are observed between the properties of the beaten films and those of the evaporated films. With evaporated films, a dependence of film properties on the rate of condensation has been noted. In view of the differences observed, which may be expected to arise from differences in film structure (which latter feature was not examined), no purpose is served by giving more than a broad indication of the results obtained on such films.

Figure 6.6 shows the results obtained by P. ROUARD, D. MALÉ and J. TROMPETTE[3] for the variation of the reflectances *R* and *R'* vs.

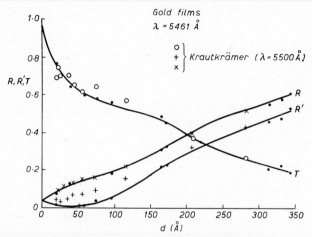

Figure 6.6   Reflectance and transmittance curves for gold films.   Full lines: measurements of Rouard, Malé and Trompette (λ = 5460 Å).   Points: results of Krautkrämer for films evaporated on quartz.   (λ = 5500 Å)

thickness for gold films deposited on glass by thermal evaporation. Films of different thicknesses were obtained by evaporation through a sectored shutter; the rates of deposition of the films therefore differ one from another. This is unfortunate since the results of R. S. SENNETT and G. D. SCOTT[4] for silver, discussed below, show clearly the marked influence of the rate of deposition on the optical properties. On the same curve are shown the results of J. KRAUTKRÄMER[5] for gold films produced by evaporation on quartz. Values of *R*, *R'* and *T* are given for a wide range of wavelengths in the visible spectrum. There is good agreement in the results shown in *Figure 6.6* except for those for *R'* at small values of the thickness.

Figure 6.7 shows results of Rouard *et al.* for a wavelength 4920 Å (obtained by interpolation) compared with early results of R. SCHULZE[6]. These measurements were made on beaten gold foils which had been thinned. Both electrolysis and sputtering were used to reduce the film thickness and no differences were found in the properties of the films produced by the two methods. The properties are probably characteristic of the metal in the state of the beaten leaf. The curves are not directly comparable with Rouard's figures since the beaten foils had no substrate backing. The reflectances for the higher thickness values will be but little affected however and the markedly higher reflectances of the evaporated films are significant. The observed differences are much

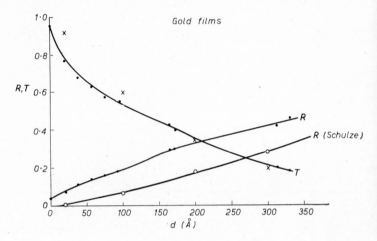

Figure 6.7 Reflectance and transmittance of evaporated gold films (interpolated for λ = 4920 Å from Rouard's results) compared with those of Schulze for beaten gold foils, electrolytically thinned

greater than are to be accounted for by the presence of the glass backing. Electron diffraction observations on gold leaf by T. B. RYMER and C. C. BUTLER[7] show anomalous lattice spacings which can be ascribed to the existence of stresses in the nature of surface tension forces. The specimens examined contained randomly oriented crystallites. Subsequent work on evaporated gold films showed a tendency to (111) preferred orientation in films of thicknesses comparable with those of the leaf specimens examined. The crystallite size was found to be smaller for the evaporated specimens. These differences in structure may well be sufficient to account for the difference in reflectances observed.

Figure 6.8 shows the variation of absorption $(A = 1 - R - T)$ exhibited by Rouard's and by Schulze's films. The values are given for $\lambda = 4920$ Å, the Rouard figure being obtained by interpolation. Also shown are results by Sennett and Scott[4] for rapidly evaporated films.   The lower absorption shown by the rapidly evaporated films is consistent with the findings of the spectroscopists who aluminize their own interferometer plates.   At high rates of deposition the proportional interference from residual gas molecules is smaller (see e.g. reference [8]).

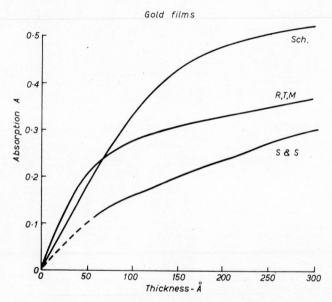

Figure 6.8   Variation of absorption with thickness.   Gold films.   R,T,M.   Rouard, Trompette & Malé.   (Evaporated films.   Interpolated to $\lambda = 4920$ Å.)   Sch. Schulze. (Thinned beaten foil. $\lambda = 4920$ Å.)   S. & S.   Sennett & Scott.   (Rapidly evaporated films (2 seconds). $\lambda = 5000$ Å)

From observations on the reflectance and transmittance of beaten gold films Schulze[6] concluded that there is no variation of optical constants with thickness for thicknesses up to 20 Å.   (In the light of later results by P. L. CLEGG[9], discussed below, this is unlikely to apply to evaporated films.)   The observed reflectance and transmittance of the films agrees roughly with the value obtained by calculation from the bulk optical constants.   It has been noted in connection with certain of the methods for determining optical constants that more than one pair of n, k values may yield the same R, T values. In the case of the gold films it is found that the observed values of the phase change on reflection at the surface differ widely from those calculated from the bulk constants.

Silver films

This metal has been extensively studied, particularly by Sennett and Scott[4] who have made a detailed examination of the optical properties of evaporated films together with a study of the structure of the films by electron microscopy. The effect of varying the rate of condensation of the films was also studied and the observed differences in optical behaviour correlated with the changes in

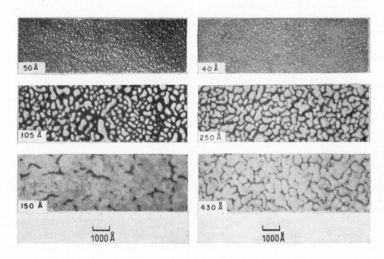

Figure 6.9   Electron micrographs of evaporated silver films

structure consequent upon the changes in the evaporation conditions. The properties of rapidly evaporated silver films have also been studied by R. C. FAUST[10].

The marked dependence of crystal structure on rate of condensation is shown in *Figure 6.9*. The left-hand micrographs are of films evaporated in about two seconds; those on the right are of films prepared by slow evaporation, lasting ten minutes. The much more granular structure of the slowly deposited films is apparent. *Figure 6.10* shows the influence of the rate of deposition on the curves of reflectance and transmittance vs. thickness of the films for a wavelength 6500 Å. Beyond a thickness of about 150 Å, the absorptions in the rapidly condensed films are much lower than those of the slowly deposited layers. Faust's results for a wavelength of 6300 Å

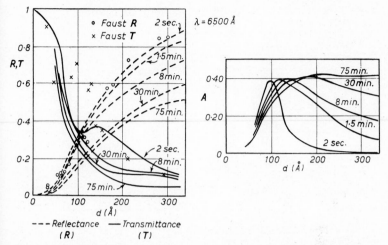

Figure 6.10 Reflectance and transmittance curves for silver films prepared at different rates of deposition ($\lambda = 6500$ Å)

Figure 6.11 Variation of absorption with thickness of silver films for various evaporation rates ($\lambda = 6500$ Å)

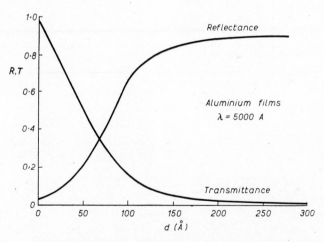

Figure 6.12 Reflectance and transmittance of aluminium films ($\lambda = 5000$ Å)

167

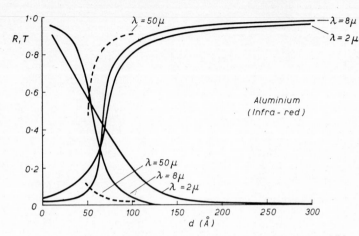

Figure 6.13   Results for aluminium films in the infra-red (2–50 μ)

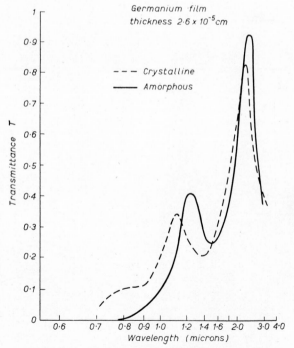

Figure 6.14   The infra-red transmission of germanium films

and a high (unspecified) rate of deposition are also shown on this curve. *Figure 6.11* shows the dependence of absorption on the rate of deposition. It is significant that 150 Å is the thickness at which the rapidly deposited films approximate in structure to bulk silver in that the granularity characteristic of thinner or of slowly deposited films has disappeared at this thickness.

Sennett and Scott have examined the absorption vs. thickness relation for several metals besides silver. In certain cases maxima appear in the A vs. d curve. This behaviour is shown to be consistent with the prediction of the Maxwell Garnett theory; the results are discussed in this connection in Section 6.8 below. As mentioned above, the position of the minimum of the R' vs. d curve varies with wavelength for silver films. Selective reflection occurs, therefore, when white light is incident on a silver or gold film from the glass side and the film appears coloured. P. ROUARD[11] gives the following tables for the colours of thin Ag and Au films on glass. The films here referred to were prepared by sputtering.

Table 6.2

Silver Films			Gold Films		
Light yellow	.	10 Å	Ruddy purple		15 Å
Golden yellow	.	21 Å	Indigo	. .	20 Å
Orange yellow	.	32 Å	Blue	. .	27 Å
Ruddy orange	.	43 Å	Green	. .	32 Å
Crimson	.	52 Å	Yellow-green	.	40 Å
Indigo	.	60 Å	(Golden yellow >40 Å)		
(colours disappear at 70 Å)					

(The film must be deposited on a wedge and not on a parallel-sided plate in order that reflection from the uncoated face may be eliminated.)

Other metals

Work on other metals has been less extensive generally than on the materials discussed above. The reflectance and transmittance of aluminium films prepared by evaporation on fluorspar have been determined for several wavelengths up to 8 μ by W. WALKENHORST[12] and in the further infra-red (20–90 μ) by W. WOLTERSDORFF[13]. No details of the rate of condensation are given but by comparison of the reflectance and transmittance curves for a wavelength of 5000 Å (*Figure 6.12*) with those of Sennett and Scott for silver

(*Figure 6.9*), it seems likely that the rate of evaporation was fairly high. The results of infra-red measurements on aluminium are shown in *Figure 6.13*. The variation of *R*, *T* with wavelength from 20–90 μ is extremely small. Woltersdorff has also determined the reflectance and transmittance of films of antimony and bismuth in this wavelength region. As would be expected from the electrical properties of these materials, the absorption is much lower in this region than that for metals.

The results of A. H. Gebbie[32] on germanium films bring out clearly the marked dependence of the optical properties of the film on the state of order. When germanium is condensed on a cold substrate, the deposit is practically amorphous. Electron diffraction patterns indicate a crystallite size of not more than 10–15 Å. When such a film is annealed at 525° C for an hour or so, crystallization becomes highly developed, as is shown by the very sharp diffraction rings obtained. The optical transmission of amorphous and crystalline films is shown in *Figure 6.14*. The interference maxima and minima show the low absorption of germanium in the near infra-red and the shift in the position of the maxima on annealing demonstrate the considerable change in the real part of the complex index. A less marked change is also observed in the extinction coefficient.

6.3. The Phase Changes on Reflection at and Transmission by a Single Film

The difficulties of making absolute reflectance measurements have been mentioned in Chapter 5. The polarimetric method of examining surfaces and films is capable of much higher accuracy than the photometric method and has been extensively used in thin film investigations. The quantities determined by the polarimetric method are (i) the ratio of the amplitudes of the two components of the reflected light, viz. that polarized in the plane of incidence to that polarized perpendicular to this plane and (ii) the difference in phase between these components.

The main disadvantage of the polarimetric method would seem to be that non-normal incidence is used and that, in the event of preferred orientation of crystallites in the film, application of the theory which assumes homogeneity may lead to senseless results. The view that normal incidence measurements (which, even with preferred orientation present, may not yield impossible results) are any more satisfactory is, however, hardly to be taken seriously. Certain workers have nevertheless assumed that when the derived

170

value of the film thickness turns out to be complex, the real part of the quantity obtained gives the film thickness. If the theory of homogeneous layers be applied to non-homogeneous layers, then the results are not expected to be meaningful even if they are real. In fact, polarimetric observations have been successfully used to determine the inhomogeneity of a surface. (See Section 6.9.)

Experimental results

Considerable attention has been paid to the determination of the phase change on reflection at films of silver for various wavelengths. This is of the utmost importance for work with the Fabry-Pérot etalon (or with any interferometer employing silvered reflecting

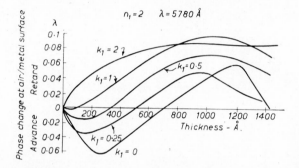

Figure 6.15 Computed curves of phase change (at air surface) vs. thickness of a film with $n_1 = 2$ for various values of k_1. The wavelength used is 5780 Å and the substrate index is 1·5

surfaces which is used for high precision measurements) since the optical separation of the plates includes the path equivalents of the phase changes at the silver surfaces. Since the phase change on reflection at a silver film is found to depend on wavelength, the effective thickness of the etalon is therefore a function of the wavelength.

(i) *Phase change on reflection at the air-film surface.*—The form of the variation of the phase change on reflection ($\Delta_{0p,s}$) with film thickness may be obtained from equation 4(136) for an absorbing film. *Figure 6.15* shows the curves obtained for a film for which $n_1 = 2$ for various values of k_1, for the case of normal incidence. In sharp contrast to the curve obtained by putting for n_1, k_1 the values appropriate to bulk silver, shown in *Figure 6.16*, are the experimental results obtained from measurements on evaporated silver films,

171

shown in the same graph. Besides showing that the optical constants of the film differ from those of bulk silver, these observations show further that the optical constants of the film depend markedly on its thickness. It is found impossible to obtain a fit between the experimental results and curves calculated from equation 4(136) for any (constant) values of n_1 and k_1, over the thickness range considered. Subsequent measurements of the values of n_1 and k_1 for films show a very considerable variation of these quantities with

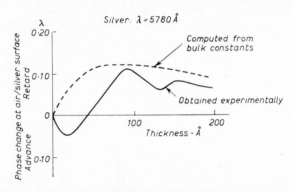

Figure 6.16   Experimental results for silver films compared with results computed from the constants of the bulk metal. $(\boldsymbol{n} = 0\cdot18 - 3\cdot67i)$

thickness. These results are discussed in Section 6.4 below. The points on *Figure 6.16* are values of Δ_0 calculated from equation 4(136) using the values of n_1 and k_1 obtained from reflectance measurements on films of the thicknesses indicated. The agreement between the Δ_0 values calculated in this way and the directly measured values shows that, although the film is known not to be homogeneous, it may be regarded as the equivalent optically of a layer possessing an effective index n and extinction coefficient k.

Using a photo-electric method of analysing the reflected light, P. L. Clegg[9] has determined the differential phase change $(\Delta = \Delta_{0p} - \Delta_{0s})$ and the value of tan ψ $(= \varrho_{0p}/\varrho_{0s})$ for evaporated silver films at various wavelengths in the visible region of the spectrum. The results for $\lambda = 5461$ Å and an evaporation rate of 2 Å per second are shown in *Figure 6.17*. Also shown in these figures are the results of D. G. AVERY[14] and the curve computed from the bulk values of the optical constants. The three sets of results are in good agreement for film thicknesses above about 200 Å. Reference to the electron micrographs of Sennett and Scott (*Figure 6.9*) shows that, for thicknesses greater than this value, the films show little granularity for the indicated rate of evaporation. The departure from the theoretical curve at thicknesses below 100 Å is violent. It is shown that this behaviour is to be expected from films; when the aggregated nature of the film is taken into account, theory predicts that the real part (n) of the index rises to a high value and the extinction coefficient (k) falls to a low value as the film thickness assumes a very low value.

The results obtained for silver are similar to those exhibited by films of indium, tin and gold. For longer wavelengths a positive

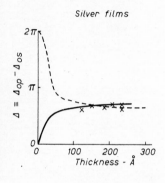

Figure 6.17 Differential phase change on reflection at evaporated silver films. Full line: Clegg's results. Dotted line: theoretical curve. Crosses: Avery's results

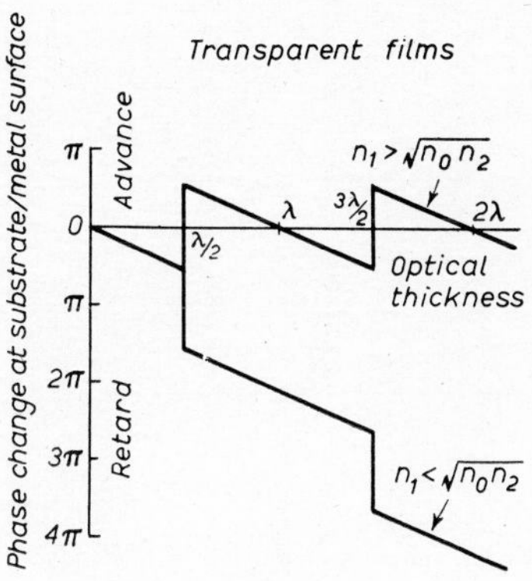

Figure 6.18 Phase change at substrate/metal surface on reflection at a transparent film. (Computed)

phase change may occur for small thicknesses, changing to a negative value for a thickness generally below 200 Å. For shorter wavelengths a retardation is shown at all thicknesses.

(ii) *Phase change on reflection at the substrate-film surface.*—The variation of phase change on reflection at the substrate side of the film, for the case of a transparent substrate (the case of an absorbing substrate is of vanishing interest), shows interesting and striking characteristics. The value is readily found from equation 4(136) by replacing n_0 by n_2 and vice versa. Considering firstly the transparent layer, we find that the march of the phase change vs. thickness curve depends on the relative sizes of n_1 and $(n_0 n_2)^{\frac{1}{2}}$. A discontinuous jump in the phase change of π occurs at optical thicknesses corresponding to $\lambda/2, 3\lambda/2, 5\lambda/2 \ldots$.

For an absorbing film, the form of the phase change curve to be expected depends largely on the value of the extinction coefficient. For feebly absorbing materials, a retardation in phase is to be expected at all thicknesses. For heavily absorbing materials, an advance is predicted. At intermediate values, an abrupt jump in the phase change, from a retard to an advance is obtained, as is shown in *Figure 6.19*. The change from a retardation at all thicknesses to the case in which an advance occurs at small thicknesses, with the accompanying abrupt phase change, takes place for a value

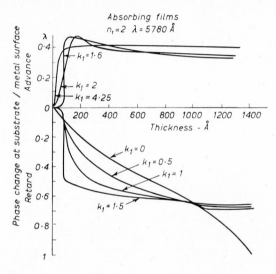

Figure 6.19   Phase change at substrate / metal surface on reflection at absorbing films (Computed)

of k_1 between 1·5 and 1·6, with $n_1 = 2$ and $\lambda = 5780$ Å. This raises the interesting possibility that, for a material in which the extinction coefficient is a function of wavelength, we may observe a continuous retardation with thickness for some wavelengths and the characteristic jump from a retardation to an advance for other wavelengths. The measurements of Rouard[11] show that this behaviour is observed for films of gold, silver and platinum. *Figure 6.20* gives the results obtained with films of platinum and shows in addition the curve of phase change vs. thickness calculated from the optical constants corresponding to bulk platinum. The phase change on reflection is seen, when these results are compared with those shown in *Figure 6.5*, to be a much more sensitive indication of the variation

174

of optical constants than are the reflectance or transmittance of the film.

(iii) *Phase change on transmission through the film.*—Equation 4(142) gives the value of the phase change γ on traversing the film, a quantity of importance in connection with the phase contrast microscope. In this instrument the necessary phase difference between the first order and zero order spectra is conveniently inserted

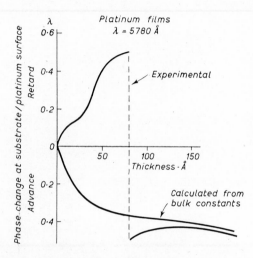

Figure 6.20 *Rouard's measurements on sputtered platinum films compared with curve computed from bulk constants*

by placing in the light path a thin film giving the appropriate phase change to the transmitted light. Since the zero-order and first order beams differ considerably in amplitude, an absorbing material is sometimes used for the phase plate in order to lessen this difference and so improve contrast. The total phase change of a light beam on traversing an absorbing layer in which multiple reflections may be ignored is conveniently dealt with in a piecemeal fashion, as shown by Rouard[11]. From equation 4(45) we see that the transmitted amplitude may be written

$$T = \tau e^{i\gamma} = \frac{(1+r_1)(1+r_2)e^{i\delta_1}}{r_1 r_2 + e^{2i\delta_1}} \qquad \text{.... } 6(6)$$

The factors $(1+r_1)$, $(1+r_2)$ include the changes in phase at the first and second surfaces of the film respectively. Let these phase

175

changes be ψ_1, ψ_2. The term $e^{i\delta_1}$ gives the phase change on traversing the film. At normal incidence this will be equivalent to a phase change of $2\pi n_1 d_1 / \lambda$. The denominator gives the effect of multiple reflections within the film. For metal films of thickness greater than about 300–400 Å, the amplitudes of multiply-reflected beams become negligibly small and the contribution of this term to the phase change becomes negligible. For films of such thickness, the total phase change in the beam on traversing the film becomes

$$\gamma \doteqdot \psi_1 + \psi_2 + 2\pi n_1 d_1 / \lambda \qquad \text{.... } 6(7)$$

We note in passing that, in measuring the phase difference introduced by the beam we shall generally compare the phase of the beam passing through the film with that which does not traverse the film but which instead traverses a thickness d_1 of medium of index n_0. The observed phase difference will therefore be

$$\gamma_{obs} = \psi_1 + \psi_2 + \frac{2\pi(n_1 - n_0)d_1}{\lambda} \qquad \text{.... } 6(8)$$

Thus for thicknesses such that the effect of multiple reflections may be neglected, the phase change on transmission increases linearly with film thickness. Since ψ_1, ψ_2 are fixed, the sign of γ_{obs} may be positive or negative, for sufficiently large values of d_1, depending on whether n_1 is greater or less than n_0.

6.4. The Optical Constants of Thin Films

The brilliant colours observed when thin films of many metals are viewed by reflected light show in a striking manner the critical dependence of optical constants on film thickness and hence on film structure. Thus whilst the reflectivity of bulk silver in the visible region of the spectrum varies but slowly with wavelength, indicating a slow change of optical constants with wavelength, silver films of thicknesses up to about 60 Å show brilliant colours, ranging from bright yellow at a thickness of about 10 Å to deep blue at about 60 Å. These thicknesses are much too low for interference effects to play any part in determining the colour reflected. The highly selective reflection occurring at such films indicates a rapid variation of the extinction coefficient with wavelength, with resultant rapid variation of reflectance. The properties of gold films are found to be similar to those described above for silver with the further feature that the colours have been observed to depend considerably on the method of preparation of the films. (References [5] and [15-17].) Thus films formed by evaporation appear light yellow when viewed by

176

transmitted light for a thickness around 10 Å whereas films of this thickness prepared by sputtering on glass appear violet. If the sputtering takes place in hydrogen and the film is formed on quartz, 10 Å films appear grey. In spite of some uncertainty in the thickness estimates in this region, there is undoubtedly a real difference in the properties of the films produced under the different conditions. Further evidence for the change of optical constants with film thickness is provided by the discontinuous jump in the phase change on reflection at the film, described in Section 6.3 above.

Maxwell Garnett theory

The optical properties of a system consisting of a large number of spherical metal particles, of diameter small compared with the wavelength of the light used and embedded in a dielectric medium, were investigated in 1904–6 by J. C. MAXWELL GARNETT[18]. In the light of the results of structure examinations of thin films, it is clear that the thin film approximates more nearly to the system considered by Maxwell Garnett than to a parallel-sided, homogeneous slab of bulk metal. The proposed model, envisaging spherical particles possessing properties identical with those of the bulk metal, is a considerable simplification of the real thing; the agreement between the predictions of the simple theory and the experimental results, although in some instances not quantitatively impressive, is more than enough to demonstrate the essential correctness of this approach to the problem. The theory has received considerable development in the hands of H. SCHOPPER[19], some of whose results are discussed below.

It is assumed that the film may be represented by spherical metal crystallites, of diameter small compared with the film dimensions. The distribution in space of the particles is assumed random and the density may be characterized by the factor q giving the fractional volume occupied by the metal. Within a sphere of radius large compared with the mean separation between crystallites, there will be a large number of particles. Provided q is not too small, the radius of such a sphere will be small compared with the film dimensions. (In practice, values of q observed are generally greater than 0·5.)

When an electromagnetic wave traverses such a system, the spherical crystallites will be polarized by the electric field E of the wave and will thus produce an additional field E_s. This field may be calculated in precisely the way in which the polarization field in a dielectric is determined. The field contributed by the dipoles may be divided into two parts; (a) that from a spherical region, around the point in question, of radius large enough to contain many dipoles and (b) that from the remainder of the medium. If the diameter of the sphere is small compared with the light wavelength used, so that the value of E at any instant does not vary appreciably over the sphere, then the contribution from (a) is zero. The field due to material outside the sphere is $\frac{4\pi}{3}P$ where P is the electric moment per unit volume of the medium. The total field F is therefore given by

$$F = E + E_s = E + \frac{4\pi}{3}P \qquad \text{.... } 6(9)$$

Since the distribution of spheres is random, the medium is isotropic and we may write $P = kNE$ where k is a constant and N the number of crystallites per unit volume. Writing $F = \varepsilon E$, we thus obtain

$$\frac{\varepsilon - 1}{\varepsilon + 2} = \frac{4\pi}{3}kN \qquad \text{.... } 6(10)$$

the well-known Clausius-Mosotti relation. Thus $(\varepsilon - 1)/(\varepsilon + 2)$ is proportional to the density ϱ of crystallites in the medium. Setting $\varepsilon = n^2$, we may write

$$\frac{n^2 - 1}{n^2 + 2} \cdot \frac{1}{\varrho} = \text{constant} \qquad \text{.... } 6(11)$$

The thin film may be represented by a volume fraction q of spheres of index n in a medium of refractive index sensibly equal to unity. The effective index n_e of the system is then given by

$$\frac{n_e^2 - 1}{n_e^2 + 2} = q\frac{n^2 - 1}{n^2 + 2} \qquad \text{.... } 6(12)$$

so that the effective index of a layer formed of spheres of given index n may be determined (with some labour) for various values of q.

The value inserted for n is generally that appropriate to the bulk material. This will certainly not be expected to be correct for systems corresponding to those observed in thin metal films for the following reasons. The crystallite sizes observed in films in the thickness range where aggregation occurs and where the 'anomalous' optical constants are exhibited are of the order tens of ångström units. Particles of metal of this size would not be expected to show bulk optical behaviour, even if they were structurally perfect crystals, on account of their size alone. Where the dimensions of the particle are small compared with the mean free path of the conduction

178

electrons, a diminution in conductivity results from surface scattering with consequent decrease in the effective optical constants. The mean free path observed for the conduction electrons in metals is of the order a few hundred ångström units.

The assumptions on which the use of the Lorentz-Lorenz polarization term $4\pi P/3$ are based are not very well justified for the thinnest of films, for which the film thickness is certainly not large compared

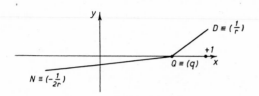

Figure 6.21 Malé's method for labourless determination of effective optical constants as a function of the volume factor q

with the crystallite size. Nor indeed can the crystallites be accurately represented as spheres. Although they are of no particularly symmetrical shape, there is evidence that for some materials the assumption of an ellipsoidal particle gives better agreement with the experimental results. (See Section 6.5 below).

Malé's method for calculating n_e [20]

If equation 6(12) is expanded and the real and imaginary parts of n_e expressed in terms of n and k, the expressions obtained are intolerably cumbersome. The following simple graphical method enables n_e and k_e to be obtained with little labour. Equation 6(12) may be written as

$$n_e^2 = \frac{1 + 2rq}{1 - rq} \quad \text{where} \quad r = \frac{n^2 - 1}{n^2 + 1} \quad \text{.... 6(13)}$$

r is calculated in the form $Ae^{i\alpha}$, where

$$A = \left\{ \frac{(n^2 - k^2 - 1)^2 + 4n^2k^2}{(n^2 - k^2 + 2)^2 + 4n^2k^2} \right\}^{\frac{1}{2}}$$

and arc tan α = arc tan $\left(\dfrac{2nk}{n^2 - k^2 - 1} \right)$ − arc tan $\left(\dfrac{2nk}{n^2 - k^2 + 2} \right)$

and n_e is written in the form

$$n_e = \left\{ -\frac{2(q + 1/2r)}{(q - 1/r)} \right\}^{\frac{1}{2}} \equiv Be^{i\beta}$$

179

B and β are readily found by plotting in the complex plane the points $\mathcal{N}\left(-\dfrac{1}{2r}\right)$, $D\left(\dfrac{1}{r}\right)$ and $Q(q)$. (*Figure 6.21*). B, β are then simply given by

$$B=\left\{2\frac{Q\mathcal{N}}{QD}\right\}^{\frac{1}{2}} \quad \text{and} \quad \beta=\frac{\widehat{DQ\mathcal{N}}+\pi}{2} \qquad \text{.... 6(14)}$$

so that $\boldsymbol{n}_e = n_e - ik_e$ is found from

$$n_e = B \cos \beta$$
$$k_e = -B \sin \beta \qquad \text{.... 6(15)}$$

The dependence of n_e and k_e on the volume factor q is striking (*Figure 6.22*). The effective values of one or other of the film constants may be higher than those of the crystallite material. In

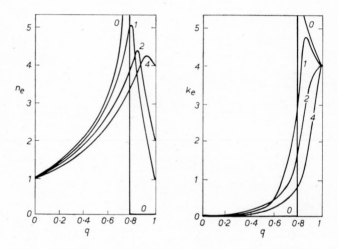

Figure 6.22   Variation of n_e and k_e with volume factor q.   Constants of bulk material are
$$\boldsymbol{n} = n - 4i \text{ where } n = 0, 1, 2, 4$$

particular, the real part n_e of the index may rise considerably, especially for small values of n. Large values of k are seen to give rise to very low values of k_e with diminishing values of q. Thus films of materials with typically metallic values of n and k—viz. low n values and high k values—are expected to show non-metallic properties (high n_e, low k_e) in thin film form. This behaviour is observed for several metals; the results are discussed below.

6.5. RESULTS OF MEASUREMENTS ON GOLD FILMS

Gold has been studied more extensively than any other metal, possibly on account of the ease with which films may be prepared, together with the absence of any tendency to oxidation. Even before the techniques of sputtering or thermal evaporation were developed it was possible by beating to produce gold films of sufficiently small thickness to possess high transparencies in the visible region of the spectrum. This point is of considerable interest in that there exists the possibility of studying beaten films and

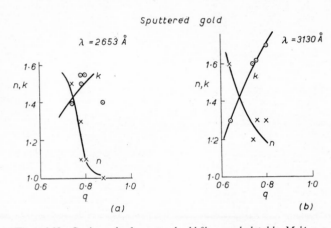

Figure 6.23 Goos's results for sputtered gold films, recalculated by Malé

evaporated films of the same thickness so that differences in bulk and film properties should be directly observable. In addition to the observations of the optical behaviour of gold films, there have been many electron diffraction examinations of the structure of gold layers, both of gold leaf and of films formed by evaporation and sputtering. The various observations may be collated to form a reasonably clear picture.

Early measurements

The optical constants of gold films, prepared by sputtering, were determined by F. Goos[15]. Measurements of R, R' and T were combined with thickness determinations by weighing. It was assumed that the density of the films was the same as that of bulk gold. Using the method described in Section 5.7, Malé has recalculated Goos's results. Since the Malé treatment yields the values of n and k simultaneously with the film thickness d, the factor q may be determined by comparing the thickness found optically with the value determined from the film mass. The recalculated results show tendencies similar to those found by subsequent investigators for evaporated films, namely that the factor q decreases with decreasing thickness, that the value of n increases and that of k decreases as the factor q decreases (*Figures 6.23 (a)* and (*b*)). These trends are to be expected in the light of the results of the Maxwell Garnett theory discussed in the last section. The values of n and k are seen (*Figure 6.24*) to approach those obtained for bulk gold as the film thickness increases, although for the thickest films ($\sim$500 Å)

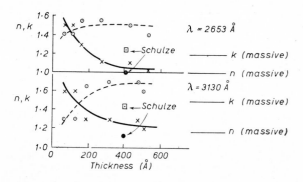

Figure 6.24   Goos (Malé) and Schulze's values for the optical constants of gold.   (Goos: sputtered films.   Schulze: thinned gold foil)

examined by Goos, the volume factor q is appreciably less than unity. The agreement is thus in a sense fortuitous especially as no details are available of the structures of the films examined. Also shown on *Figure 6.24* are the results obtained by R. Schulze[6] for gold leaf thinned electrolytically to 400 Å. The results are much closer to those of W. Meier[21] for bulk gold than to the thin-film values and suggest that the q-value for the beaten foils is likely to be unity.

More recently, D. Malé and P. Rouard[22] have applied the method evolved by the former author to the results obtained on evaporated gold films. The authors state that, in view of the dependence of film properties on the conditions of preparation, which are usually not stated, a further set of measurements on gold was desirable. Films deposited under carefully controlled conditions at a rate of deposition of 60–80 Å per minute were examined. Although the q-values for these films decreases steadily with decreasing thickness, the values of n and k vary somewhat erratically and do not tend towards the bulk values of Meier as their thickness increases. Also the q-values are found to vary appreciably for different wavelengths although there seems no obvious reason why they should do so.

Krautkrämer's measurements[5]

An extensive investigation on gold films produced by evaporation is reported by Krautkrämer. The structures of films were examined by electron diffraction. The optical constants and resistivity of the films were determined and the influence on these factors of variation of the substrate (quartz) temperature was studied.

The thickness was determined from the mass of metal volatilized during an evaporation, a preliminary experiment serving to indicate that the method could be relied upon to within about 5 per cent. In the trial experiments, for which thick films ($\sim$200 Å) were used, the values of k obtained by Goos were used, it being assumed that the variation of k with thickness above 200 Å could be neglected. This is probably justified to within the ± 5 per cent quoted. The optical constants were determined from the measured values of **R**, **R'** and **T** together with the thickness value. Typical results obtained for gold films condensed on quartz at 20° C for three wavelengths are shown in *Figures 6.25* (a) and (b). The variations of n and k with thickness are most violent for films less than about 100 Å thick, for a substrate temperature of 20° C. The reversal in the values of both n and k is observed. The variations with wavelength of the n and k values for the four thinnest films and for the thickest film (281 Å) are shown in *Figure 6.26*. These curves may be compared with those of *Figure 6.27* showing the variation to be expected on the Maxwell Garnett theory for values of q from 0·5–1·0. A rough similarity in shape is seen between the experimental results and those calculated for $q = 0·6$–0·7. Close agreement is not, however, to be expected since the n, k values have been determined from the mass-per-unit-area thickness figure d_w. In the event of the film possessing a q value less than unity, the optically determined thickness will be greater than d_w and the n, k values will need to be redetermined. The final values can clearly only be obtained by a trial-and-error process and this illustrates forcibly the disadvantage of using a method in which the thickness is not determined simultaneously with the optical constants. *Figure 6.28* shows the agreement between the theoretical n, k vs. wavelength curves for a film of (true) thickness 88 Å compared with the observed value, assuming a q-value of 0·6. The difference between the two curves does not arise from a poor

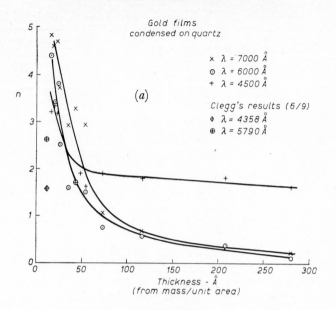

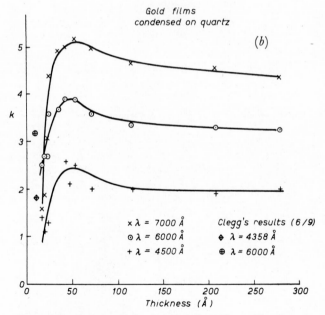

Figure 6.25   Optical constants of evaporated gold films.   (Krautkrämer)

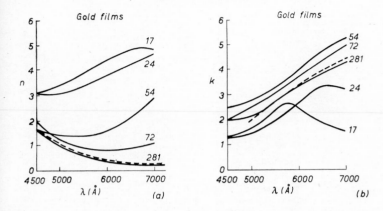

Figure 6.26 Variation of n, k with wavelength (Krautkrämer). Film thickness (Å) shown on each curve. The dotted curves show the results of Haringhuizen et al. for thick films

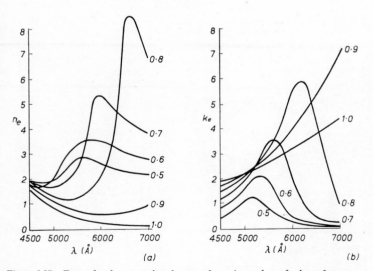

Figure 6.27 Form of n, k vs. wavelength curves for various values of volume factor q

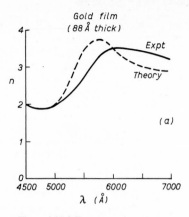

Figure 6.28 *Experimental results for a film of thickness 88 Å compared with those calculated for a film with volume factor 0·6*

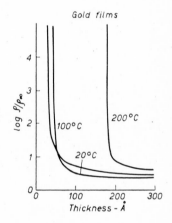

Figure 6.30 *Dependence of resistivity of gold films on substrate temperature at deposition*

Figure 6.29 *Dependence of n, k on substrate temperature at deposition*

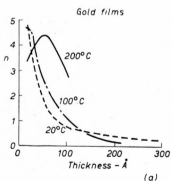

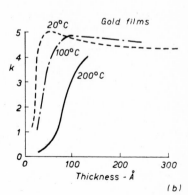

choice of q-value. In the light of Schopper's results discussed below, it is likely that this difference springs from the use of an invalid assumption in calculating the results using the Maxwell Garnett theory, viz. that the gold particles in the film are spherical.

The dependence on substrate temperature of the n and k results are summarized in *Figure 6.29.* The maxima in the n, k vs. thickness curves move in the direction of greater thicknesses as the substrate temperature is raised. The electron diffraction patterns taken indicate a larger crystal size in the layers deposited at the higher temperatures, together with a tendency to (111) orientation of the crystallites. The larger crystal size results from the higher mobility of the atoms on the substrate surface (see Chapter 2).

Measurements of the resistivity of the films deposited at different substrate temperatures (*Figure 6.30*) show that the critical thickness, below which the film resistivity becomes enormous, increases with increase of substrate temperature. This is to be expected in the light of the observed tendency to form larger aggregates at higher temperatures. The film becomes conducting only when there are sufficient crystallites in contact with one another. Thus a thin layer, consisting of small crystallites in contact will conduct. The same amount of metal deposited at a higher substrate temperature forms a smaller number of larger crystallites which do not now make contact with one another. (A slightly naïve picture, but qualitatively correct.)

Further work on Krautkrämer's results shows that the degree of fit with the Maxwell Garnett picture is better for films deposited at the higher than at the lower temperature, for thicknesses up to 100–150 Å. The reasons for poorer agreement with the films deposited at 20° C are not hard to find. The fact that films thicker than about 50 Å (*Figure 6.30*) have an appreciable conductance shows that the simplified Maxwell Garnett model of isolated spheres in a non-conducting medium is not fulfilled. Furthermore, at such small thicknesses, the crystallite size is so small (as shown by the very diffuse electron diffraction patterns obtained) that the assumption that the metal spheres behave as though possessing the bulk optical constants will be very wide of the mark. The films deposited at 200° C contain much larger crystallites, for which this latter assumption, although still certainly not correct, will be less in error. Also the fact that high conductivity in such films is not attained until the thickness reaches about 200 Å, indicates that these films are a decidedly closer approximation to the Maxwell Garnett picture.

187

Schopper's measurements

The Maxwell Garnett theory has been developed into its most sophisticated form to date by Schopper[19]. Gold films have formed the object of application of the development of the theory. The reasons for the want of fit between the Krautkrämer results and the model based on spherical crystallites becomes apparent and considerably closer agreement is attained by the use of a model based on a two-dimensional distribution of particles in the shape of rotation ellipsoids.

The films examined, ranging in thickness from about 25 to 250 Å

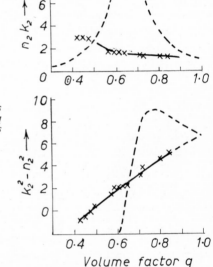

Figure 6.31 Experimental values of $n_2 k_2$ and $k_2^2 - n_2^2$ vs. q compared with calculated values assuming spherical particles

Volume factor q

were deposited by evaporation on to quartz and the values of n, k and d were determined by measuring the reflectances at each side of the film and the transmittance. Both amplitude and phase were measured, as described in Section 5.7. The masses of the films were also measured, by direct weighing, so that the volume factor q could be directly determined for each film. The optical constants to be expected on the Maxwell Garnett model, in which a spherical particle shape is assumed, were then calculated and comparison made with the experimental results. The latter make a quite uncompromising departure from the theoretical curves as shown in *Figure 6.31*. It will be recalled that in Krautkrämer's results, in which experimental and theoretical expressions for n_e, k_e vs. q were considered, it was found impossible to get a close fit by adjustment of the value of q. (The greater departure in Schopper's results is apparent only, resulting from the use of $n_2 k_2$ and $(k_2^2 - n_2^2)$ instead of simply n_2 and k_2.) It has been mentioned already that real thin films are but an approximation to the Maxwell Garnett picture. We cannot generally consider a sphere surrounding a given crystallite in the film as containing a very large number of metal spheres, as we require to do in order to apply the Lorentz-Lorenz argument. Most of such a sphere would in fact lie outside the film. Schopper, following early work on these lines by E. DAVID [23], assumes that the film can be represented by a two-dimensional distribution of particles. The particles are assumed to be ellipsoids of rotation with the unique axis lying perpendicular to the substrate. It is further assumed that the axial ratios of the ellipsoids are distributed statistically about a mean value. There is support from two directions for the idea of flat, plate-like crystallites in thin films. Study of the electron micrographs of Sennett and Scott shown in *Figure 6.9* reveals that for e.g. a rapidly evaporated silver film 110 Å thick, the average 'diameter' of the aggregates across the film surface is several hundred ångström units. Furthermore, the anomalies in the radii of electron diffraction rings observed by T. B. RYMER and C. C. BUTLER [24] in gold specimens suggest that the film consists of flat platelets lying with their planes parallel to the mean plane of the film.

Exact calculation for a system such as is envisaged above is extremely hard. Within the restriction that terms in $(d/\lambda)^2$ may be neglected compared with unity—with which films showing the 'anomalous' behaviour amply comply—the variation of the optical constants to be expected with the dimensions of the constituent ellipsoids may be obtained.

Consider initially the simple case of a layer of ellipsoidal particles of the same size and axial ratio. The average film thickness is d; that deduced from the mass per unit area on the assumption of bulk density d_w. If n_e and k_e are the observed values of the optical constants for such a system, and n and k the constants of the bulk metal, we have

$$\frac{d}{d_w}(n_e^2 - k_e^2 - 1) - i \cdot \frac{d}{d_w} \cdot 2 n_e k_e = C \qquad \ldots.. \; 6(16)$$

$$\text{where } C = \frac{(n - ik)^2 - 1}{[(n - ik)^2 - 1]f + 1}.$$

189

f is a calculable function of the axial ratio (b/a) of the ellipsoids whose variation with b/a is seen in *Figure 6.32*. $(n_e^2 - k_e^2 - 1)d$ and $2n_e k_e d$ are readily obtained from the amplitude and phase respectively of the light reflected by such a film. Crystallite sizes in a real film are, however, continuously distributed about a mean value so the assumption of a constant value for the axial ratio b/a of the ellipsoids is a trifle too simple. The obvious

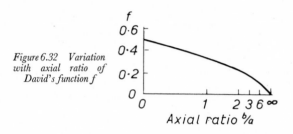

Figure 6.32 Variation with axial ratio of David's function f

elaboration is to assume that the function f of the axial ratio is distributed in Gaussian fashion about a mean value $\bar{f}$ where the distribution function for f is

$$g(f) \propto f \exp\left(-f/\bar{f}\right)^2 \qquad \text{.... } 6(17)$$

C in equation 6(16) above is then replaced by the mean value $\bar{C}$ defined by

$$\bar{C} = \frac{\int Cg(f)df}{\int g(f)df} \qquad \text{.... } 6(18)$$

The values of n_e and k_e to be expected on this model may therefore be calculated as a function of $\bar{f}$, instead of as a function of q, the volume factor, as in the simpler theory.

Figures 6.33 (a) and *(b)* show the computed curves and superposed on them the experimental results obtained by Schopper on evaporated gold films. On the whole the agreement is strikingly good. The thinnest films (11 Å) behave as though the mean axial ratio of the ellipsoids were about 1:8 and this ratio decreases as the film thickness increases. For a slab of bulk material, $d = d_w$ and $g(f) = 0$. Under this condition $C = (n - ik)^2 - 1$ and from equation 6(16) we see that $n_e = n$ and $k_e = k$.

Homogeneity of gold films

Although the indications from the behaviour of gold films of increasing thickness are that their optical properties tend steadily

190

towards the values characteristic of the bulk material, there is some evidence that the bulk values may never be reached, suggesting that the structure of evaporated or sputtered layers never becomes the same as that of bulk gold. D. MALÉ[25] notes that the optical constants of evaporated and of sputtered films tend towards the same value as the thickness increases, which suggests that their structures are substantially the same. The limiting values of n and k attained at thicknesses beyond about 200 Å are

$$\lambda = 4358 \text{ Å} \qquad n = 1\cdot4, \quad k = 1\cdot8$$

$$\lambda = 5461 \text{ Å} \qquad n = 0\cdot4, \quad k = 2\cdot3$$

The volume factor for such films is found to be about 0·9.

The assumption underlying the Maxwell Garnett treatment, both in the simple form and in the more sophisticated form developed

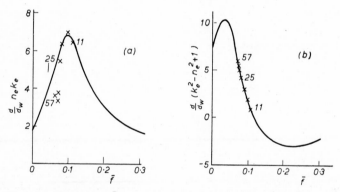

Figure 6.33   Variation with $\bar{f}$ of optical constants for model of ellipsoidal gold particles compared with experimental results.   The figures beside the points give film thicknesses in Å

by Schopper is that the structure of the film does not vary throughout its thickness. H. WOLTER[26] showed that for a film lying in air on a substrate of index n_2, the absorption for light traversing the film from the air side (A) is related to that for light travelling in the reverse sense (A') by $n_2 A = A'$. This relation forms a convenient test for the homogeneity of films. The extent to which the relation holds for the various published results varies somewhat, the homogeneity of the films plausibly depending on the conditions of preparation. Departures are greater for thicker films suggesting that the evident substrate influence on the film structure becomes small for thicker films. The departures shown by Krautkrämer's results are considerable; those of Schopper are much less marked. In view of the difficulties of allowing for the many factors which play a part in determining the optical behaviour of gold films, the measure of agreement obtained is encouraging.

6.6. RESULTS OF MEASUREMENTS ON SILVER FILMS

Measurements on silver films have been less extensive than those on gold films. There are, moreover, larger discrepancies between different workers' results than is the case with gold. Some early measurements by Goos[15], originally dealt with by Murmann's method (Section 5.7) have been recalculated by Malé. These results were obtained in the ultra-violet region of the spectrum. In the visible spectrum, extensive measurements have been made by Krautkrämer[5] and have been discussed in connection with the simple Maxwell Garnett theory, as was done by the same worker for gold films. One of the difficulties associated with Murmann's method of obtaining n and k is the fact that the curves obtained for n vs. k may intersect in two places. Both pairs of n, k values so obtained are plausible for thin films, such are the vagaries of the latter's optical constants. For silver films, one pair of values has k less than and the other has k greater than unity. From the computed curves of R, R' and T against film thickness, it is seen that a maximum occurs in the transmission vs. thickness curve at very small thicknesses $(d/\lambda \sim 0.01)$ only if k is less than unity. Such a maximum is, in fact, observed for silver films in the visible spectrum. Krautkrämer chose the solution giving $k > 1$ for films deposited on a substrate at $20°$ C, although his results for films deposited at higher temperatures give values of k less than unity over part of the wavelength range $(4000–7000 \text{ Å})$ examined. Similarly in dealing with Goos's results, the solutions with $k > 1$ have been taken. In recalculating Goos's results using the method for obtaining optical constants and thickness simultaneously, Malé finds the k values to be generally less than unity. The resultant variation of n and k with thickness is shown in *Figure 6.34.* It may be remarked that the march of the recalculated values is less erratic than was that obtained with the original values. Also shown in *Figure 6.34* are the results obtained by Goos for bulk silver. As with gold films, it appears that the optical constants of indefinitely thick evaporated films of silver do not attain those characteristic of the bulk material. The results for sputtered films are similar to those for films produced by thermal evaporation.

In view of the uncertainty of Krautkrämer's figures for silver films deposited on a substrate at room temperature, any detailed discussion is unlikely to be fruitful. Krautkrämer has applied the

Maxwell Garnett theory to the results obtained for silver films deposited at 400° C, for which k values less than unity are quoted. Substantially similar behaviour is observed as in the case of gold films condensed at room temperature. The nearest fit between observed and calculated values of n and k is obtained for a q value between 0·6 and 0·7 although in no case is the fit very close. The reasons are presumably the same as those applying to the gold layers; the model assuming spherical particles is too simple. In

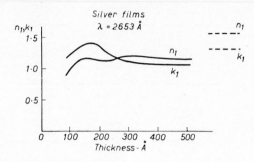

Figure 6.34   Results on silver films.   (Goos, recalculated by Malé)

view of the known inhomogeneity of the films, as shown by the considerable departure from Wolter's rule, the labour of applying Schopper's modification of the Maxwell Garnett theory would hardly be well spent.

Using a sensitive polarimetric method, Clegg[9] has determined the ellipticity of the light reflected from the vacuum side of evaporated silver films of thicknesses up to about a hundred ångström units. (The measurements were made on the films *in vacuo*, immediately after deposition.) As expected, wide divergences are shown between these values and those calculated from the bulk constants. Application of the Maxwell Garnett theory shows that the thinnest films (10 Å bulk thickness) possess a volume factor q of around 0·6. The results of measurements made with a wavelength of 5461 Å are shown in *Table 6.3*. The values for bulk silver at this wavelength are $n = 0\cdot1$, $k = 3\cdot3$. The reversal of the magnitudes of n and k expected on the Maxwell Garnett picture is clearly shown by these results. Although the fit between the experimental curves and the points calculated from the above figures is not very close, the values of n and k are probably substantially correct for rapidly evaporated silver films. Steps were taken to ensure that variations from experimental conditions were eliminated and the optical measurements were made on the films in vacuum. The values quoted above differ violently from the earlier figures obtained with silver films, among which there existed alarming discrepancies. (Values of n for 20 Å films have been variously reported as 0·5, 1·67 and 3·5 for a wavelength 5500 Å.)

Table 6.3. *Optical Constants of Evaporated Silver Films (Clegg[9])*

Thickness (bulk) Å	n	k
10	4·0	1·0
33	3·25	2·0
44	3·5	3·0
77	2·0	3·5

6.7. Results of Measurements on Other Metals

Studies of materials other than silver and gold have been less extensive than those quoted above. Clegg has examined thin (25 Å) films of tin and has analysed the results on the basis of Maxwell Garnett's (simple) theory and finds reasonable agreement. Avery has studied much thicker layers of tin and also layers of copper by the multiple-beam method described in Section 5.7, obtaining values of n and k appreciably different from those of the bulk material. O'Bryan has measured n and k for films of several metals, the films being prepared by evaporation. Since this work was done at a time when the variation of optical constants with film thickness was unsuspected, the thickness of the films was not determined. They were presumably opaque so that one can expect the results to be appropriate to an 'indefinitely thick' film, to which the results for thin layers tend as the thickness grows. As has been previously observed, these limiting values of n and k do not necessarily correspond to those of the bulk material. Gebbie has studied the optical behaviour of germanium films with special reference to their state of order.

Platinum, palladium and copper [27–29]

Measurements have been made, using Försterling's method (Section 5.7), of the values of n and k for sputtered films of the above metals. Values are given of the variation with film thickness over the range 10–200 Å, and all three materials show a similar trend, the index

n falling and the extinction coefficient k rising with increasing thicknesses up to 100–150 Å and there becoming fairly steady. The curves for these metals are given in *Figures 6.35 (a)–(c)*. Shown in *Figure 6.35 (a)* are values for bulk platinum, interpolated for this wavelength from the early results of Meier (1903). The values marked A in *Figure 6.35 (c)* are Avery's values[14] for copper films

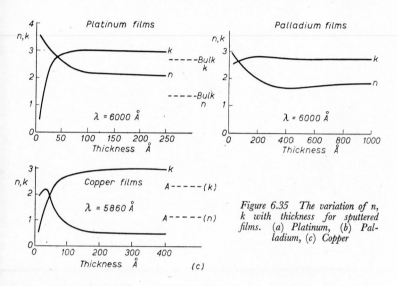

Figure 6.35 The variation of n, k with thickness for sputtered films. (a) Platinum, (b) Palladium, (c) Copper

having a transmittance at $\lambda = 5400$ Å of a few per cent. The thicknesses of these layers will be around 400–500 Å. The several quoted values for n and k for bulk copper show a marked dependence on the method of preparation, ranging from $n = 1 \cdot 0$, $k = 2 \cdot 3$ (mean values) for polished specimens to $n = 1 \cdot 38$, $k = 1 \cdot 78$ for electrolytically deposited specimens. Avery's values for evaporated films are seen to be much closer to the bulk (polished) figures than to the values obtained for sputtered films. As no examinations have been made of the crystal structures of the various specimens used, little can be said about these differences.

Germanium

With the growth of interest in the photoconductive properties of this material, attention has been turned to studies of its optical constants. The want of concordance in the results of the three investigations made on films of evaporated germanium is somewhat alarming. A redeeming feature is the emergence of definite direct evidence for the dependence of optical constants on the structure of the film.

The results of observations in the visible spectrum are collected in *Figure 6.36*. All the results shown were obtained on films of thickness above a few hundred ångströms, at which value the initial violent convulsions of the n, k vs. thickness curves may be expected to have died away. The curves of H. M. O'BRYAN[30] are for

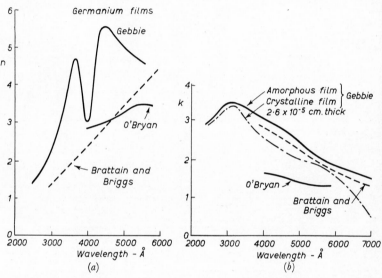

Figure 6.36   Measurements on evaporated germanium films: (a) index, (b) extinction coefficient

evaporated films of unspecified thickness. Since the purpose of the measurements was to determine values of n and k which would correspond to those of the bulk material, the films were probably opaque to visible light. In view of the low extinction coefficient of germanium at the red end of the visible spectrum, complete opacity to a white source indicates a thickness of several thousand ångström units.

The values obtained by W. H. BRATTAIN and H. B. BRIGGS[31], also for evaporated films, show wide divergences from O'Bryan's values. The optical constants were in these experiments determined from the transmittance of the films. The thicknesses of the films ranged from 300 to 1600 Å. Calculated values of the reflectance of the films from the measured optical constants showed reasonable agreement with the observed reflectances for thick films.

The reflectance calculated from O'Bryan's figures is, however, rather considerably below the observed value. There seems no obvious reason for the very large difference between these two sets of figures, particularly with regard to the sign of the difference in the extinction coefficients. One possible reason for differing results would tend to make the O'Bryan figures for *k higher* than those of Brattain and Briggs. For the films used by Brattain and Briggs were heat-treated after deposition. Electron diffraction evidence shows the tendency for large crystallites to grow during such a process and the results of Gebbie mentioned below show that heat-treatment tends on this account to reduce the value of the extinction coefficient below that of an unheated specimen.

The effect of annealing on germanium films was carefully examined by A. H. GEBBIE [32], electron diffraction studies being made of films before and after annealing. Films deposited on a cold substrate were found to be nearly amorphous and to possess resistivities enormously larger than those of the bulk source material. The resistance of the films decreased on heating and at 525° C steady values were attained only after one to two hours annealing. The electron diffraction pattern then showed very fine rings characteristic of a highly crystalline state. The films examined by Brattain and Briggs were heat-treated at the lower temperature of 400° C. Such treatment produced an appreciable change in the transmittance vs. wavelength curve, the interference maxima and minima moving in the direction of shorter wavelengths and the transmittance increasing. *Figure 6.36 (b)* shows, however, that the Brattain and Briggs curve of *k* vs. wavelength lies between the curves obtained by Gebbie before and after annealing at the higher temperature. It seems likely, therefore, that the process of annealing at the lower temperature has not resulted in as complete crystallization and that for this purpose the higher temperature is required.

Measurements of the optical constants of bulk germanium have been made by I. SIMON [33] on polycrystalline specimens and by D. G. AVERY and P. L. CLEGG [34] on a single crystal. The two sets of results differ considerably from one another, particularly the extinction coefficients. For the polycrystalline specimens, much lower values of *k* are recorded. The results obtained by Gebbie for annealed evaporated films show values of *k* nearer to those obtained with the single crystal than to those for the polycrystalline bulk. The values of *n* obtained by Simon are closer to the Brattain and Briggs figures than to those obtained by Gebbie. On the whole, there is depressingly little correspondence between the different observers' results.

Tin

Avery's measurements on evaporated tin films, using the multiple-beam fringe method, yield values of n well above those of bulk tin; the values of k for the films are well below the bulk figures. The thicknesses of these films are likely to be a few hundred ångströms. For layers 25 Å thick, Clegg obtains still higher values for n and lower k values for a wavelength 5461 Å. Comparison with values calculated from the Maxwell Garnett theory shows close agreement. The results are summarized in *Table 6.4*.

Table 6.4. *Optical Constants of Tin Films*

Nature of Film	λ	n	k
Few hundred Å (reflectance = 0·7)	5400*	2·4	1·9
25 Å (expt) . .	5461	3·0	0·5
25 Å (calc)† . .	5461	2·8	0·3
25 Å (expt) . .	3560	2·3	1·0
25 Å (calc)† . .	3560	2·2	1·5
Bulk . . .	5461	1·0	4·2

* Filter transmitting 5400 ± 150 Å.
† The q-value used in the calculation is not stated. Probably 0·6.

O'Bryan's results for Mg, Be, Ca, Ba, Sr, Al, La, Mn and Ce [30]

In addition to the measurements on germanium discussed above, O'Bryan has measured n and k for films of the above-mentioned elements. The layers were prepared either by evaporation from tungsten spirals, molybdenum crucibles or graphite crucibles. The thicknesses are not stated, but, as in the case of germanium, they must be presumed to be such that the limiting n and k values, corresponding to an indefinitely thick film, are attained. The results are summarized in *Figures 6.37 (a)–(d)*. The points marked on *Figures (c)* and *(d)* are the n and k values for bulk aluminium ($n - \odot$; $k - \triangle$) and for manganese ($n - \times$; $k - +$) and show some difference in the film constants from those of the bulk. Little can be said in this connection since the crystalline structures of the films are not known.

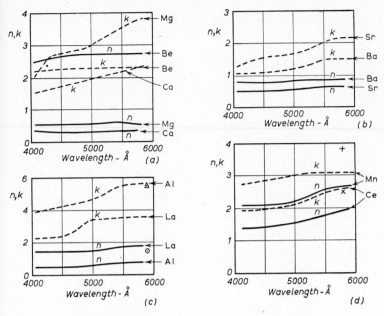

Figure 6.37   O'Bryan's results of n and k for various metals

Schulz's measurements on silver, gold, copper and aluminium

Using the methods described in Section 5.7, L. G. SCHULZ[35-36] has determined the values of n and k for thick films of the above metals, aged and annealed so that the reflectance relations at the boundaries corresponded to those of a smooth, homogeneous layer.   Good agreement with previously recorded values of k for the bulk metals was obtained whereas considerably lower values of n were found. This suggests that the films used in these determinations were free from strains and defects generally, which tend to reduce the conductivity of the film and thus increase the real part of the index. The values obtained by this method probably correspond fairly closely with those exhibited by the surface of an annealed bulk specimen (although these may still not correspond to the values effective in the interior of the crystal).   For films of tin and lead, it was not found possible to obtain the relation $R_s^2 = R_p$ by annealing, suggesting that the film method is unsuitable for the estimation of even approximate bulk n, k values for these metals.   The values of n, k for silver, gold, copper and aluminium are summarized in *Table 6.5*.

Table 6.5

Wavelength	Optical Constants							
	Silver		Gold		Copper		Aluminium	
(Å)	n	k	n	k	n	k	n	k
4000	0·075	1·93	1·45	—	0·85	—	0·40	3·92
4500	0·055	2·42	1·40	1·88	0·87	2·20	0·49	4·32
5000	0·050	2·87	0·84	1·84	0·88	2·42	0·62	4·80
5500	0·055	3·32	0·34	2·37	0·72	2·42	0·76	5·32
6000	0·060	3·75	0·23	2·97	0·17	3·07	0·97	6·00
6500	0·070	4·20	0·19	3·50	0·13	3·65	1·24	6·60
7000	0·075	4·62	0·17	3·97	0·12	4·17	1·55	7·00
7500	0·080	5·05	0·16	4·42	0·12	4·62	1·80	7·12
8000	0·090	5·45	0·16	4·84	0·12	5·07	1·99	7·05
8500	0·100	5·85	0·17	5·30	0·12	5·47	2·08	7·15
9000	0·105	6·22	0·18	5·72	0·13	5·86	1·96	7·70
9500	0·110	6·56	0·19	6·10	0·13	6·22	1·75	8·50

Pitfalls

In concluding this section we may again recall the considerable differences which are sometimes shown by the optical constants of thick evaporated films from those of the bulk material. The Maxwell Garnett theory shows clearly how the apparent optical constants depend on the structure of the film and on the q-value. It is abundantly clear that an exact quantitative treatment of the problem, which must needs take into account the effects of crystallite shape and size, of finite size of crystallite in relation to the optical constants of the particle, etc., would be rather complicated. At the beginning of this chapter it was mentioned that the ease of preparing films with optically smooth surfaces offered promise as a means of obtaining the optical constants of materials without the complicating effects of structural disorder by polishing. On the assumption that the values obtained in such measurements are characteristic of the material in bulk, conclusions were earlier drawn from these measurements about the energy-band structure of the solids examined. This is seen to be a hazardous procedure since, as shown by the Maxwell Garnett picture, variations occur in the extinction coefficient simply from differences in film structure.

6.8. ABSORPTION IN THIN FILMS

The absorption of electro-magnetic waves in a metal is due in part to free electrons and in part to bound electrons. In certain regions of the spectrum one or other of these two sources of absorption may be dominant. The contribution of free electron absorption in films consisting of aggregates will be considerably higher than that in the bulk metal, on account of the much larger extent of scattering at the particle boundaries. The contribution of the bound electrons will be expected to change but slowly with the state of aggregation. Considering the variation of absorption with film thickness, in which the variation arises from the change with thickness of the degree of aggregation, we shall expect a marked dependence of absorption on thickness for wavelengths at which the absorption is mainly by free electrons and a much slower dependence for wavelength regions where the effect of bound electrons predominates. This is precisely what is observed. Silver and aluminium show very marked maxima in their absorption vs. thickness curves for all wavelengths. Copper and gold films show similar behaviour for wavelengths above 5500 Å. Films of antimony, nickel, palladium and chromium show only slow variation of absorption with thickness for visible wavelengths as do gold and copper for wavelengths below the absorption edge at 5500 Å.

Sennett and Scott[4] have shown that the appearance or otherwise of a maximum in the absorption vs. thickness curve may be predicted in terms of the optical constants of the bulk metal with the aid of the Maxwell Garnett theory. A film of metal of optical constants n, k and volume factor q exhibits apparent optical constants n', k' related to n, k and q by the following expressions

$$n'k' = \frac{3qb}{(1 - qa)^2 + 4q^2b^2} \qquad \text{.... } 6(19)$$

$$k'^2 - n'^2 = 2 - \frac{3(1 - qa)}{(1 - qa)^2 + 4q^2b^2} \qquad \text{.... } 6(20)$$

where
$$a = \frac{(k^2 - n^2 - 1)(k^2 - n^2 - 2) + 4n^2k^2}{(k^2 - n^2 - 2)^2 + 4n^2k^2} \qquad \text{.... } 6(21)$$

and
$$b = \frac{3nk}{(k^2 - n^2 - 2)^2 + 4n^2k^2} \qquad \text{.... } 6(22)$$

Differentiation of 6(19) shows that the curve of $n'k'$ vs. q (which latter increases with film thickness) possesses a maximum if $a^2 + 4b^2 > 1$.

From 6(21) and 6(22), this corresponds to $K > 0$ where

$$K = 2(k^2 - n^2)^3 - 9(k^2 - n^2)^2 + (12 + 8n^2k^2)(k^2 - n^2) - 4 - 12n^2k^2$$
$$.... \ 6(23)$$

The signs of value of K for several metals are shown in *Table 6.6.*

Table 6.6.   (*Sign of K for various metals*)

Metal	λ	Sign of K
Aluminium	5890	+
Antimony	5890	−
Chromium	6080	−
Copper	3950	−
Copper	6000	+
Germanium	3000	+
Germanium	5000	−
Gold	4000	−
Gold	5800	+
Manganese	5790	−
Nickel	5890	−
Palladium	6000	−
Platinum	5890	−
Silver	5890	+

The variation of absorption with thickness for several metals shown in the table are seen in *Figure 6.38.*   No maxima are observed for metals for which $K < 0$, in agreement with the predictions of the Maxwell Garnett theory.

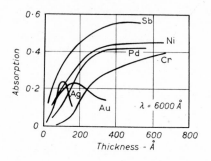

Figure 6.38   *Variation of absorption with thickness for various metals*

6.9. Optical Inhomogeneity of Films

We conclude this chapter with a reference to the optical behaviour to be expected of films in which the refractive index varies in a direction normal to the film surface. In all the developments carried out thus far it has been assumed that the films are homogeneous except on a very small scale. The theory developed for a film of constant index has been elaborated to deal with the structural

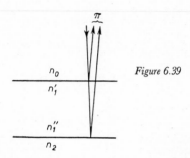

Figure 6.39

character of evaporated films but has assumed that the properties of the layer next to the glass are identical with those next to the air. It seems likely that large differences of this kind do not exist in evaporated films of single, pure materials (although no experiments seem to have been made to confirm this point). Films with such differences can be made and, by choice of the appropriate form of variation of film index with depth, films with interesting and useful properties can be obtained.

For a glass surface, of refractive index n_2 covered by a film in which the index varies continuously from n_1' on the air side to n_1'' on the substrate side (*Figure 6.39*), the reflectance of a film of such thickness that it introduces a phase difference of π between reflections at each side of the film is given by

$$R = \left(\frac{n-s}{n+s}\right)^2 \qquad \qquad \text{.... } 6(24)$$

where $s = n_1''/n_1'$.

(We note that if s could be made equal to n then such a film would be a perfect anti-reflector for all wavelengths, neglecting any differences in dispersion between the film and substrate materials. Although this condition cannot generally be realized, inhomogeneous films are more efficient anti-reflectors than single homogeneous ones. See Chapter 7.)

The reflectance of the plain glass at normal incidence is given by $(n-1)^2/(n+1)^2$ as is also the reflectance of a homogeneous film introducing a phase difference of π. Lack of homogeneity therefore manifests itself in a difference in the peak reflectance (maximum or minimum depending on the sign of $n_1 - n_2$) and that of the uncovered substrate. The difficulties of accurate reflectance measurement, however, make the measurement of inhomogeneity in terms of such a reflectance difference an insensitive and unsatisfactory method. It is far better to employ the more highly sensitive polarimetric methods.

A. Vašíček[37] has shown that the principal azimuth, measured with light of a wavelength such that the phase difference between successive beams is π, forms a useful indication of the extent of the inhomogeneity of a film. The sensitivity of the method is high; inhomogeneities in refractive index of $\sim$0·001 are detectable by this means. The choice of the conditions given above results from the fact that, at principal incidence for such a film, the principal azimuth ψ for a homogeneous film is the same as that of the uncoated substrate. The measurement of ψ is therefore effected on the uncoated substrate. The difference $(\Delta\psi)$ between this and the value obtained over the film is found to vary nearly linearly with the difference in refractive index (Δn_1) between the two sides of the film, even for appreciable differences between these values.

As with most polarimetric derivations, the one relating $\Delta\psi$ and Δn_1 is trigonometrically tedious. The broad outlines are given by Vašíček, yielding the result

$$|\Delta n_1| = \frac{n_1(n_1^2 - \sin^2\varphi_0)^{\frac{1}{2}}\Delta\psi}{\sin\varphi_0 \cos^2\psi \tan\psi \left\{ \dfrac{b(m+m')}{a+b} - \dfrac{md}{c+d} - \dfrac{ab(m+m')}{1+ab} + \dfrac{cmd}{1+cd} \right\}}$$

.... 6(25)

where φ_0 is the angle of incidence at the air side,

$$a = \frac{\tan(\varphi_0 - \varphi_1)}{\tan(\varphi_0 + \varphi_1)} \qquad\qquad b = \frac{\tan(\varphi_1 - \varphi_2)}{\tan(\varphi_1 + \varphi_2)}$$

$$c = \frac{\sin(\varphi_0 - \varphi_1)}{\sin(\varphi_0 + \varphi_1)} \qquad\qquad d = \frac{\sin(\varphi_1 - \varphi_2)}{\sin(\varphi_1 + \varphi_2)}$$

$$m = \cot(\varphi_1 - \varphi_2) - \cot(\varphi_1 + \varphi_2)$$

$$m' = \tan(\varphi_1 - \varphi_2) - \tan(\varphi_1 + \varphi_2)$$

φ_1 is the mean angle of refraction in the film and φ_2 the angle of refraction in the glass support.

204

These expressions are valid only for an inhomogeneity sufficiently small for one to be able to characterize the film by a mean index n_1 and to neglect the change in the angle of incidence of the light in the film. This is reasonably well obeyed if n_1—1 does not vary by more than some 10 per cent. For larger variations, the usual (exact) expressions for ψ' and ψ are evaluated for various inhomogeneities. We note that n_1 is required in the evaluation of Δn_1.

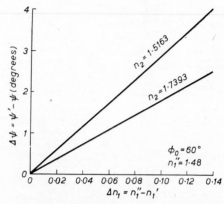

Figure 6.40   Variation of $\Delta \psi$ with film inhomogeneity

The value of the mean index of the film is readily found by any of the methods described in Section 5.6.

The relations between $\Delta\psi = \psi' - \psi$ and Δn_1 for two different substrates covered by films of various indices, giving the requisite phase difference of π at an angle of incidence of 60°, are shown in *Figure 6.40*.

The sensitivity of the method may be judged from the calculated results for films on glass of refractive index 1·5163. For an angle of incidence 60°, the values of $\Delta\psi$ corresponding to $\Delta n_1 = 0·02$ range from 20·5' for $n_1 = 1·30$ to 35·5' for $n_1 = 1·70$. Since polarimetric methods enable azimuths to be determined readily to within 1', it is seen that quite small inhomogeneities are detectable.

References

[1] HAMMER, K.   *Z. Techn. Phys.* **24** (1943) 169
[2] PARTZSCH, A. and HALLWACHS, W.   *Ann. der Phys.* **41** (1913) 247
[3] ROUARD, P., MALÉ, D. and TROMPETTE, J.   *J. de Phys.* **14** (1953) 587
[4] SENNETT, R. S. and SCOTT, G. D.   *J.O.S.A.* **40** (1950) 203
[5] KRAUTKRÄMER, J.   *Ann. der Phys.* **32** (1938) 537

[6] Schulze, R.   *Phys. Zeit.* **34** (1933) 24

[7] Rymer, T. B. and Butler, C. C.   *Proc. Phys. Soc.* **59** (1947) 541

[8] Crawford, M. F., Gray, W. M., Schawlow, A. L. and Kelly, F. M. *J.O.S.A.* **39** (1949) 888

[9] Clegg, P. L.   *Proc. Phys. Soc.* **65** (1952) 774

[10] Faust, R. C.   *Phil. Mag.* **41** (1950) 1238

[11] Rouard, P.   *Propriétés Optiques des Lames Minces Solides*, LIV, Gauthier-Villars, Paris

[12] Walkenhorst, W.   *Z. Techn. Phys.* **22** (1941) 14

[13] Woltersdorff, W.   *Z. Phys.* **91** (1934) 230

[14] Avery, D. G.   *Phil. Mag.* **41** (1950) 1018

[15] Goos, F.   *Z. Phys.* **106** (1937) 606

[16] Pogany, B.   *Ann. der Phys.* **49** (1916) 531

[17] Haringhuizen, P. J., Was, D. A. and Druithof, A. M.   *Physica* **4** (1937) 695

[18] Maxwell Garnett, J. C.   *Phil. Trans.* **203** (1904) 385; *ibid.* **205** (1906) 237

[19] Schopper, H.   *Z. Phys.* **130** (1951) 565

[20] Malé, D.   *C. R. Acad. Sci. Paris* **230** (1950) 286

[21] Meier, W.   *Ann. der Phys.* **31** (1910) 1017

[22] Malé, D. and Rouard, P.   *J. de Phys.* **14** (1953) 584

[23] David, E.   *Z. Phys.* **114** (1939) 389

[24] Rymer, T. B. and Butler, C. C.   *Proc. Phys. Soc.* **54** (1947) 541

[25] Malé, D.   Thèse, Université de Paris, 1952

[26] Wolter, H.   *Z. Phys.* **105**, (1937) 269

[27] Pogany, B.   *Ann. der Phys.* **49** (1916) 531

[28] Goldschmidt, H. and Dember, H.   *Z. Techn. Phys.* **7** (1926) 137

[29] Planck, W.   *Phys. Z.* **15** (1914) 563

[30] O'Bryan, H. M.   *J.O.S.A.* **26** (1936) 122

[31] Brattain, W. H. and Briggs, H. B.   *Phys. Rev.* **75** (1949) 1705

[32] Gebbie, A. H.   Ph.D. Thesis, Reading, 1952

[33] Simon, I.   *J.O.S.A.* **41** (1951) 730

[34] Avery, D. G. and Clegg, P. L.   *Proc. Phys. Soc.* B **66** (1953) 512

[35] Schulz, L. G.   *J.O.S.A.* **44** (1954) 357

[36] —— and Tangherlini, F. R. *Ibid.* 362

[37] Vašíček, A.   *J. de Phys.* **11** (1950) 346

7

PRACTICAL APPLICATIONS OF THIN FILMS IN OPTICS

*Introduction—Anti-reflecting systems—High-efficiency reflecting systems
—All-dielectric high-reflecting films in interferometry—Interference
filters—The frustrated total reflection filter—Comparison of multilayer
filters with other narrow pass-band filters—The use of thin films as
polarizers—Methods of monitoring film thickness—Miscellaneous
applications of thin films*

7.1. INTRODUCTION

IN the field of optics, use has been made of partly-transparent, highly-reflecting metal films for a considerable time. The Fabry-Pérot interferometer, first used over fifty years ago, showed in a striking manner the potential importance of thin films for optical studies. The silver films employed at that time were deposited chemically and then burnished. Subsequently, the method of sputtering was found to be more convenient. The more recently developed technique of thermal evaporation has enabled high-reflecting metal films, with low light absorption, to be deposited with ease under conditions which may be readily controlled. Much more recently, attention has been turned to the use of films of dielectric materials with striking results. From a humble, but nevertheless important and useful, beginning in which a single dielectric layer was applied to a surface to reduce the light reflected at that surface, systems of dielectric films of alarming complexity, possessing remarkable properties, have been developed. Progress in the methods of depositing such films and of controlling their thickness has enabled the properties of such combinations to be exploited rapidly. Within a very short time, the construction of complicated multilayer systems has passed from the stage of being a highly specialized laboratory technique to that of being a routine industrial process. It will be seen from the ensuing sections that in some instances, multiple-dielectric films may with considerable advantage replace the semi-transparent metal films used hitherto.

Details of the methods of monitoring the thickness of films will be given at the end of this chapter, since they are in certain cases more readily understood when the properties of the multilayer systems are first established.

7.2. ANTI-REFLECTING SYSTEMS

When light is incident on a surface separating media of differing refractive index, part of the light is reflected. Electro-magnetic theory enables the quantity reflected to be calculated on the assumption of a mathematically sharp discontinuity in refractive index. Although real boundaries are not mathematically sharp, the change in refractive index takes place over a distance which is very small compared with the wavelength of light used and under this condition the theoretical treatment gives the correct result to within very close limits. The light so reflected can be of considerable embarrassment in an optical instrument both on the grounds that it is lost from the main beam, so that any final image is weakened and also because some of the light so reflected turns up at the final image plane in the wrong place. Contrast of the image is thereby impaired.

In 1892, H. D. Taylor noticed that the light transmission of certain of his microscope objectives increased with time, and in 1916 Kollmorgen achieved a similar result by subjecting surfaces to chemical attack. Little further attention was paid to this phenomenon until the techniques for film deposition under controlled conditions had been established. With surprising rapidity, the stage has now been reached when many optical components are customarily supplied with their surfaces treated to reduce the unwanted reflection.

The reflection coefficient at normal incidence at a surface of refractive index n in air is given by $(n-1)^2/(n+1)^2$ and amounts to several per cent for a glass surface. If it were possible to coat the surface with a film whose refractive index varied continuously from n to unity (in a distance not small compared with the wavelength of the light used) then zero reflectance would be obtained at all wavelengths. This ideal cannot be realized in practice since no transparent solid materials are known with refractive indices sufficiently low. The early observations referred to above were made on surfaces on which a surface layer of lower index than the bulk material had formed. Some further work has been done along these lines, a low-index layer being produced by removal of the heavier elements (e.g. Pb, Ba) in dense glasses, so that a silicate framework, of low refractive index, remains on the surface. Glasses exposed to intense ion beams are found to possess anti-reflecting coatings, probably through a similar mechanism. These methods are, however, not easy to control and have largely given place to multilayer techniques in which the desired anti-reflecting properties are achieved by the deposition by thermal evaporation of films of suitable thickness and refractive index.

(i) *The single film*

The use of a single layer of low index as a means of suppressing the reflection from a surface is well known. At the wavelength for which the optical thickness of the film is one quarter-wavelength, the beams reflected from the upper and lower surfaces of the film differ in phase by π. From equation 4(44) we see that the reflectance at normal incidence of the film-covered surface under this condition is zero if $r_1 = r_2$. Inserting the values of r_1 and r_2, using equations 4(50) and 4(51), we obtain $n_1^2 = n_0 n_2$ as the required relation between the refractive indices. For blooming the surface of a glass of refractive index 1·50, a film of refractive index 1·22 is, therefore, required. Herein lies a difficulty. The refractive indices of known solid materials which are suitable for blooming layers (i.e. are non-absorbing, stable, strongly adhering and abrasion-resisting) are all higher than 1·22. Magnesium fluoride, with a refractive index variously reported as 1·38 to 1·40, has been used; its mechanical properties are good, especially if the film is baked after deposition. When used with a dense flint glass, almost perfect blooming (at one wavelength) is obtained. With a light crown glass, the reflectance is reduced from 4 per cent to about 1·4 per cent. A film index of about 1·23–1·28 may be formed by evaporating calcium fluoride at a pressure above 10^{-4} mm Hg. Although by this procedure the required low index is obtained, the resultant film is not very satisfactory. The mechanical properties are poor and the index tends to wander with time. Thus F. ABELÈS[1] reports a change from 1·258 to 1·278 within two days of deposition of such a film. Cryolite films, with an index 1·30–1·31 give a more effective anti-reflecting film than do magnesium fluoride films, although they are less resistant to abrasion. A drift of $\sim$0·01 within the first few days has been observed with these films. The stability of films of lithium fluoride $(n = 1·36–1·37)$ and aluminium fluoride $(n = 1·38–1·39)$ has been found to be slightly better than that of cryolite or calcium fluoride.

It is clear that a single layer as an anti-reflecting film has severe limitations. In addition to the fact that no really suitable material exists for blooming crown glass, there is the marked wavelength-dependence of the film's anti-reflecting property. Even if the film index is correct, blooming is perfect only at one wavelength. The curve of reflectance vs. wavelength in the visible spectrum for a film of magnesium fluoride ($n_1 = 1 \cdot 38$) on a glass of index $1 \cdot 52$ is shown in *Figure 7.1*. Although this degree of suppression of the reflectance is sufficient for many purposes, more effective measures are readily available by the use of two or more films.

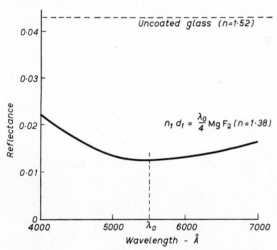

Figure 7.1   Reflectance of a glass of index 1·52 when covered with a single film of magnesium fluoride of optical thickness $\lambda_0/4$, where $\lambda_0 = 5500$ Å

(ii) *The double film*

The above discussion of the single film shows that although in principle zero reflectance is attainable (at one wavelength), the lack of suitable materials makes this result impossible to achieve satisfactorily. A system of two layers, for which there are four independent variables (two film thicknesses and two refractive indices) may in principle yield zero reflectance at two wavelengths. By suitable placing of the wavelengths at which zero reflectance occurs, the reflectance over the whole of the visible spectrum may be reduced to below the minimum reflectance obtainable with a single film. The values of refractive indices required for two-layer blooming in the visible spectrum are within the range covered by practicable materials. Although the lack of a continuous range of refractive indices makes it impossible to place the zero reflectances arbitrarily, the values available do enable very efficient two-layer anti-reflecting films to be made. In common with the terminology used in lens design, two-layer systems with two zeroes of reflectance are termed achromatic.

The properties of double layers may be easily understood from an *approximate* vector treatment which ignores multiple reflections within the films. This approximation is tolerable so long as the refractive indices employed are not too high (see Section 4.10). The vector sum of the amplitudes of the beams reflected at the

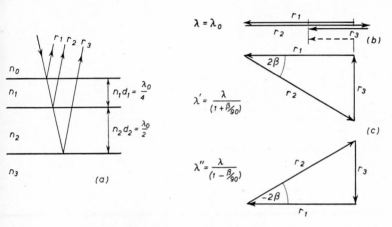

Figure 7.2 (a) Two-layer anti-reflecting system; (b) Vector diagram for wavelength λ_0; (c) Vector diagram for neighbouring wavelengths at which perfect blooming occurs

three interfaces of the system is determined graphically. Zero reflectance occurs at the wavelengths for which the vectors form a triangle. For the system shown in *Figure 7.2 (a)*, in which the upper film is of optical thickness $\lambda_0/4$ and the lower film $\lambda_0/2$, the reflected beams differ in phase successively by $180°$ and the vector diagram of *Figure 7.2 (b)* results. There are then two neighbouring wavelengths at which the vector diagrams close, as shown in *Figure 7.2 (c)*. A typical reflectance curve for a two-layer anti-reflecting film, calculated by A. F. TURNER[2], is given in *Figure 7.3*. Such curves are most conveniently plotted against phase angle $\delta = \dfrac{\lambda_0}{\lambda}.90°$ since they may then be used for any central wavelength λ_0. With λ_0 centred at $\lambda = 5500$ Å, the reflectance of the system given in *Figure*

7.3 is seen to be less than about 1 per cent over the whole of the visible spectrum. The reflectance of the uncoated glass ($n_3 = 1.62$) is 5·6 per cent.

The phase angles at which the reflectance falls to zero are determined by the refractive indices of the layers. Turner has calculated

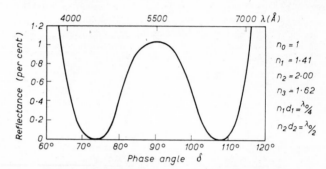

Figure 7.3   Calculated reflectance curve for a two-layer system

the refractive indices required to yield zeroes at angles of 12·5°, 15° and 17·5° on either side of the central wavelength, for the system $\dfrac{\lambda_0}{4} + \dfrac{\lambda_0}{2}$ on glass for three different substrate indices. The results are shown in *Table 7.1*.

Table 7.1.   Film Indices for a two-layer anti-reflecting system

$$\lambda'(90° + \beta) = \lambda_0 \cdot 90° = \lambda''(90° - \beta)$$

n_3	$\beta°$	n_1	n_2
1·52	12·5	1·28	1·74
	15	1·31	1·78
	17·5	1·35	1·83
1·62	12·5	1·33	1·90
	15	1·36	1·94
	17·5	1·41	2·00
1·72	12·5	1·38	2·06
	15	1·41	2·10
	17·5	1·48	2·19

Although the range of refractive indices of suitable thin film materials is limited, it is seen that certain of the above combinations

212

are realizable with existing materials. (Cryolite: 1·30–1·33; LiF: 1·36; AlF$_3$: 1·38; MgF$_2$: 1·38–1·40; CaSiO$_3$: 1·69; Al$_2$O$_3$: 1·76; AgCl: 2·06; PbCl$_2$: 2·2.)

(iii) *Triple films*

The use of three layers enables three zeroes of reflectance to be obtained. If optical thicknesses of 1, 2 and 3 quarter wavelengths are used (the $3\lambda_0/4$ layer being next to the surface), then a zero occurs at a wavelength λ_0, with two zeroes symmetrically disposed about λ_0. In general, with an N-layer system with layers of thickness 1, 2, 3......$N \times \lambda_0/4$, the reflectance curve is symmetrical

Table 7.2. Film Indices for a
three-layer anti-reflecting system

$$\lambda'(90° + \beta) = \lambda_0 . 90° = \lambda'' (90° - \beta)$$

n_4	$\beta°$	n_1	n_2	n_3
1·52	10	1·31	2·06	1·62
	15	1·33	2·03	1·64
	20	1·41	2·12	1·72
1·62	10	1·37	2·32	1·75
	15	1·39	2·27	1·77
	20	1·49	2·41	1·90
1·72	10	1·43	2·57	1·87
	15	1·46	2·53	1·91
	20	1·59	2·73	2·08

about a wavelength λ_0, and possesses N zeroes. A three-layer system, with the appropriate vector diagrams, is shown in *Figure 7.4*. Turner's calculations of the values of refractive indices required to give zeroes of reflectance at phase angles of 10°, 15° and 20° from the central zero are given in *Table 7.2*.

We see from the above table that a light flint glass ($n = 1·62$) may be effectively bloomed by a three-layer system of MgF$_2$ (1·38), ZnS (2·30) and Al$_2$O$_3$ (1·76). The high temperature required for the evaporation of Al$_2$O$_3$ introduces a slight difficulty although not an insuperable one. Since this is the first layer to be applied to the glass, there is no danger of spoiling the other layers of the system. The reflectance curve to be expected from such a system is shown in *Figure 7.5*. The side minima are not quite zero since the refractive indices of the materials do not fulfil the theoretical requirements exactly. The reflectance over a range covering almost the whole visible spectrum is reduced to below 0·05 per cent. Still lower reflectance may be obtained by using a larger number of films. It is unlikely that more efficient blooming than that provided by a triple layer would normally be required. Thus whilst for an instrument with 20 unbloomed surfaces, the proportion of light transmitted is only 32 per cent, the use of the three-layer anti-reflecting system described above increases this figure to 98 per cent for the visible

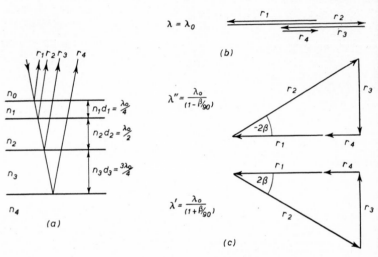

Figure 7.4   (a) Three-layer anti-reflecting system;   (b)–(c) Vector diagrams for three-layer system at wavelengths at which zero reflectance occurs

spectrum. Exact calculation of the properties of multilayer anti-reflecting systems is straightforward though laborious. Approximations which yield results of adequate accuracy have been derived by A. VAŠÍČEK[3].

Variation of reflectance with angle of incidence

From practical considerations it is important that the anti-reflecting property of the system employed should not vary rapidly with angle of incidence. The properties of the single $\lambda/4$ layer are satisfactory in this respect. The variation of reflectance with angle of incidence at the wavelength for which perfect blooming occurs is shown in *Figure 7.6*. It is seen that the increase of reflectance with angle of incidence is negligible for angles of incidence up to 50°. Exact calculations for double and triple films are somewhat laborious. Approximate methods indicate that the permissible tolerance in angle of incidence decreases with increasing number of layers and that for a triple film, the reflectance begins to rise appreciably at an angle of incidence of about 25° for light polarized parallel to the plane of incidence. For the perpendicular component, much less variation with angle of incidence is found.

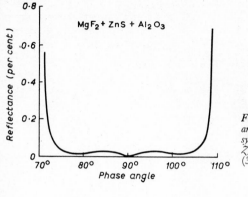

Figure 7.5 Calculated reflectance curve for three-layer system using MgF_2 ($\lambda_0/4$), ZnS ($\lambda_0/2$) and Al_2O_3 ($3\lambda_0/4$) on glass of refractive index 1·62

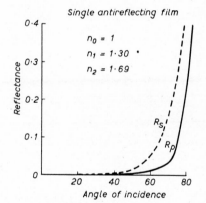

Figure 7.6 Variation of reflectance of single anti-reflecting film with angle of incidence

7.3. HIGH-EFFICIENCY REFLECTING SYSTEMS

(a) *All-dielectric systems*

In experiments requiring highly reflecting surfaces, extensive use has in the past been made of metals. The high reflectivity of silver (∼96 per cent in the visible spectrum) has made this metal an obvious choice for many purposes. Where resistance to atmospheric corrosion has been required, aluminium or rhodium have been used.

215

These metals have the further advantage that their reflectivity remains high in the ultra-violet region of the spectrum; the reflectivity of silver falls to a few per cent in the neighbourhood of $\lambda = 3200$ Å. In high resolution interferometry, e.g. using the Fabry-Pérot interferometer, the limit of resolution attainable is set by the absorption of light by the reflecting films, as well as by their limited reflectivity. No advantage can be taken of the increased reflectivity resulting from an increase in thickness of the reflecting metal film if the amount of light transmitted by the system becomes insufficient for the fringes to be seen. This is an inescapable difficulty with metal films. Even the simple task of dividing a beam into equal parts by a half-silvered mirror results in considerable loss of light by absorption in the film. Values for a silver film at $\lambda = 7000$ Å are $R = T = 0.36$, $A = 0.28$.

The need for highly-reflecting systems with very small absorption has been met in a remarkable fashion by the use of systems of dielectric layers. Such systems give a high reflectance over certain wavelength regions only, and low reflectance in others. In certain circumstances, this may possess advantages (see Section 7.4).

(i) *The single film.*—The reflectance of a surface may be considerably enhanced by the deposition of a film of material of high refractive index of optical thickness $\lambda/4$ at the wavelength at which the high reflectance is required. The refractive index of the film must exceed that of the substrate. Under this condition, the beams reflected from the air/film and film/substrate interfaces are in phase. The higher the refractive index of the film, the greater is its reflectance. Dielectric materials with refractive indices up to ~ 2.6 in the visible spectrum and possessing negligible absorption are available. In the near infra-red, semi-conducting layers of refractive index up to about 5 and with negligible absorption form very suitable high-reflecting materials for this region.

The reflectance at wavelength λ of a surface of refractive index n_2 covered by a $\lambda/4$ film of index n_1 is given by

$$R = \left(\frac{n_0 n_2 - n_1^2}{n_0 n_2 + n_1^2}\right)^2 \qquad \dots 7(1)$$

Values of reflectance obtained from single layers of common materials on a surface of refractive index 1.50 are shown in *Table 7.3*. The refractive indices are given to one decimal place only since the values obtained in practice are to some extent dependent on the conditions of preparation. For the wavelengths given, the absorptions of the materials mentioned above are very small, amounting to

216

less than 3–4 per cent for quarter-wave thicknesses. The loss in thermally evaporated films is in some cases the result of scattering by the granular structure of the films rather than by true absorption.

High-reflecting films are most conveniently deposited by thermal evaporation on account of the ease with which the required thickness may be deposited. Methods for the control of thickness are

Table 7.3. Reflectance of Surface of Index 1·50 when covered by various λ/4 Films

Film material	Refractive index	Wavelength	Reflectance of single film
ZnS	2·30	5461 Å	0·31
TiO_2	2·6	5461 Å	0·40
Sb_2S_3	2·7	1 μ	0·43
Ge	4·0	2 μ	0·69
Te	5·0	4 μ	0·79

given in Section 7.9. Chemical methods have been used to a limited extent for the production of suitable high-reflecting layers. Thus M. BANNING [4] has made films of TiO_2 by exposing the surface to be treated to the vapour from $TiCl_4$. Both TiO_2 and Fe_2O_3 films have been prepared by oxidizing thermally evaporated films of the metals, which latter are more easily evaporated than the oxides. The advances in thermal evaporation techniques have now enabled the full advantages to be taken of the method of depositing directly the required films.

(ii) *Multiple layers.*—The high reflectance of a single quarter-wave high-index layer results from the phase agreement of the beams reflected at the two interfaces. In a stack of quarter-wave films alternately of high and low index, the reflected beams from all the interfaces are in phase on leaving the uppermost boundary. With such a stack of dielectric films, reflectances very close to one hundred per cent and with negligible light-loss are to be expected on theoretical grounds and are realized in practice. The usual arrangement is as shown in *Figure 7.7*, with high-index layers next to the substrate and at the air interface.

In view of the great importance of this particular arrangement of films, the theoretical behaviour of such a system has been thoroughly investigated. The methods given in Chapter 4 may be used to determine the reflectance and transmittance of finite numbers of layers. Somewhat more simply, the properties of an infinite stack of low-high quarter-wave pairs may be determined. This has been done by A. F. TURNER[5] and is of importance in showing the limiting properties of any practical system. It is found that the theoretical properties of even a fairly small number of low-high pairs approach those of the infinite stack reasonably closely. Thus *Figure 7.8* shows the computed values of reflectance vs. phase angle for an infinite

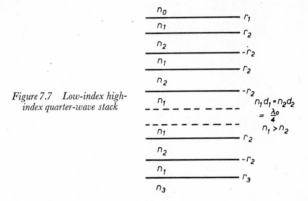

Figure 7.7 Low-index high-index quarter-wave stack

series of LH quarter-wave pairs compared with that to be expected from the system LHLHLHL (where L, H refer to layers of equal optical thickness $\lambda/4$) embedded in a medium of index higher than that of the low-index layer. The values of refractive indices used are 2·30 (ZnS) for H and 1·38 (MgF$_2$) for L.

Reflectance at the maximum.—The reflectance at normal incidence of a stack of non-absorbing layers of alternating refractive index and of optical thickness equal to $\lambda_0/4$ may be easily calculated by the method of Section 4.8. When such a stack is used as a reflecting system in an interferometer for work at one particular wavelength, this is the only quantity which is required. For a set of k layers, the matrix product of equation 4(103) reduces to

$$\begin{pmatrix} a_k & b_k \\ c_k & d_k \end{pmatrix} = (i)^k \begin{pmatrix} 1 & r_1 \\ r_1 & 1 \end{pmatrix} \begin{pmatrix} 1+r_2^2 & -2r_2 \\ -2r_2 & 1+r_2^2 \end{pmatrix}^{\frac{k-1}{2}} \begin{pmatrix} 1 & r_3 \\ -r_3 & 1 \end{pmatrix} \quad \dots \; 7(2)$$

where r_1, r_2 and r_3 are the Fresnel reflection coefficients at the boundaries n_0/n_1, n_1/n_2 and n_1/n_3 respectively. The transmittance of the system at normal incidence is calculated from equation 4(106b). The elements a_k of the product matrix are given by

$$\left.\begin{aligned}
|a_3| &= Au + Bv \\
|a_5| &= Au^2 + 2Buv + Av^2 \\
|a_7| &= Au^3 + 3Bu^2v + 3Auv^2 + Bv^3 \\
|a_9| &= Au^4 + 4Bu^3v + 6Au^2v^2 + 4Buv^3 + Av^4
\end{aligned}\right\} \quad \ldots \ 7(3)$$

etc.

where
$$A \equiv 1 - r_1 r_3 \qquad\qquad B \equiv r_3 - r_1$$
$$u \equiv 1 + r_2^2 \qquad\qquad v \equiv 2r_2$$

The value of a_k (for k odd) is readily written down for any k. The transmittance is given by

$$T_k = \frac{n_3}{n_0} \frac{(1 + r_1)^2 (1 - r_2^2)^{k-1} (1 + r_3)^2}{|a_k|^2} \qquad \ldots \ 7(4)$$

and the reflectance is found from $R_k + T_k = 1$.

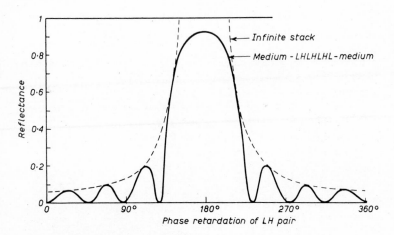

Figure 7.8   Reflectance of LH quarter-wave stacks.   Dotted line: reflectance of infinite stack. Full line: reflectance of system LHLHLHL sandwiched by high index media

The high reflectances attainable with quite a small number of layers is illustrated by the results in *Table 7.4* for two typical systems. The values of refractive indices used in calculating the reflectances are as follows:

Zinc sulphide   2·30 ⎫
Cryolite   1·30 ⎬ on glass, $n_3 = 1\cdot50$

Antimony sulphide   2·70 ⎫
Calcium fluoride   1·28 ⎬ on quartz, $n_3 = 1\cdot45$

The values for ZnS + cryolite are for sodium light: for the stibnite + CaF_2, the refractive indices at a wavelength of 1 micron are given.

Table 7.4

No. of layers				Reflectance at normal incidence	
				ZnS + cryolite ($\lambda = 5893$ Å)	$Sb_2S_3 + CaF_2$ ($\lambda = 1\ \mu$)
3	.	.	.	.	0·695	0·835
5	.	.	.	.	0·891	0·900
7	.	.	.	.	0·964	0·977
9	.	.	.	.	0·988	0·995
11	.	.	.	.	0·998	0·999

J. RING and W. L. WILCOCK[6] show that such high values are realizable experimentally and quote as typical values for a 7-layer cryolite + zinc sulphide stack at $\lambda = 4300$ Å, $\boldsymbol{R} = 0\cdot983$, $\boldsymbol{T} = 0\cdot012$, $\boldsymbol{A} = 0\cdot005$. The reflectance of the ZnS + cryolite system at $\lambda_0 = 4500$ Å is higher than that calculated for the sodium D-lines, presumably due to the appreciably higher index of ZnS at the shorter wavelength.

The use of multilayer high-reflecting stacks of this type is further discussed in Section 7.4 below in which their application to interference filters is considered.

Reflectance at any wavelength λ.—Expressions have been given by D. C. FRAU[7] for the transmittance of a system of any number of LH pairs for any wavelength. It will not be expected that such an expression will be simple. Writing

$$\delta = \pi\lambda_0/\lambda, \quad a_1 = \frac{1}{4}\cdot\frac{n_0}{n_3}\left(\frac{n_3}{n_0} - 1\right)^2, \quad a_2 = \frac{1}{4}\frac{n_0}{n_3}\left(\frac{n_3^2}{n_0^2} - 1\right)$$

$$\varepsilon = (n_2 - n_1)^2/4n_1n_2$$

$$\tan \Phi = \left\{1 - \frac{1}{[1 - 2(1 + \varepsilon)\ \sin^2\ \delta]^2}\right\}^{\frac{1}{2}}$$

220

$$g_1(\varepsilon, \delta) = \frac{1}{(\varepsilon \tan^2 \delta - 1)} \left\{ \frac{\varepsilon a_1}{\cos^2 \delta} + a_1(1 + \varepsilon) \cos^2 \delta + 1 + \varepsilon - 2a_2\varepsilon^{\frac{1}{2}}(1 + \varepsilon)^{\frac{1}{2}} \right\}$$

$$g_2(\varepsilon, d) = \frac{1}{(\varepsilon \tan^2 \delta - 1)^{\frac{1}{2}}} \{ \tfrac{1}{2}a_1(1 + \varepsilon)^{\frac{1}{2}} \sin 2\delta - a_2\varepsilon^{\frac{1}{2}} \tan \delta \}$$

then the transmittance T is given by

$$T = [(1 + a_1 \sin^2 \delta) \cosh^2 \varPhi + g_1(\varepsilon, \delta) \sinh^2 \varPhi + g_2(\varepsilon, \delta) \sinh 2\varPhi]^{-1}$$

.... 7(5)

The value of T given is the ratio of the intensity of the light measured in the substrate medium n_3 for unit incident energy. The reflectance R is thus obtained simply by subtracting from unity the value of T given by equation 7(5). Curves of T vs. wavelength are given by Frau and are similar to that shown in *Figure 7.8*. It is also shown that, by arranging two such stacks with their central reflectance maxima suitably displaced, a filter with a high transmission (> 95 per cent) and fairly narrow bandwidth may be produced. For stacks of four layers, a steep-sided band 250 Å wide centred at $\lambda = 5600$ Å and with a peak-transmission of 97 per cent is typical of the band characteristic obtainable by this method. On either side of the band the transmission remains below ~ 4 per cent down to $\lambda = 4700$ Å and up to $\lambda = 6800$ Å.

(b) *Metal + dielectric systems*

The all-dielectric high-reflecting systems described above behave as highly selective reflectors with the added advantage that their absorption is low. An alternative method of producing selective reflection, for cases in which no transmitted beam is required, is to

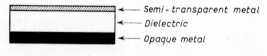

Figure 7.9

use combinations of metal and dielectric films. Systems of this type have been examined by L. N. Hadley and D. M. Dennison[8], B. H. Billings[9] and Turner[2]. An opaque metal reflecting film is covered by a dielectric film and by a semi-transparent metal film (*Figure 7.9*). The thicknesses of the two latter films are adjusted so that the beams reflected from the two metal surfaces leave the system

221

with phase agreement.   Thus the optical thickness of the dielectric layer plus the path equivalent of phase changes at the metal/ dielectric interfaces is made equal to $\lambda_0/2$ where λ_0 is the wavelength at which high reflectance is required.   By suitable choice of the resistance of the semi-transparent layer, the reflectance of the system may be made zero at wavelengths for which the total optical thickness of the system is an odd number of quarter-wavelengths.   At wavelengths for which the optical thickness is an even number of quarter-wavelengths, the reflectance of the filter is the same as that of the metal backing.   The required resistance, 377 ohms per square, may be obtained with films of aluminium or rhodium.   The characteristics of a typical filter consisting of an aluminium backing, a magnesium fluoride film to give a second-order reflectance peak at $\lambda = 6300$ Å and a semi-transparent aluminium layer are shown in *Figure 7.10*.   The transmittance of the semi-transparent layer is 35 per cent.   The selectivity of the filter may be enhanced by adding further dielectric + metal (DM) pairs.   The reflectance at the central wavelength remains the same as that of the opaque metal but falls more steeply on each side of the maximum.   Typical values for the width of the reflectance peak at half intensity (in terms of phase angle) obtained experimentally are

System	Half-width of Second Order Maximum
DM	$0 \cdot 76\pi$
DMDM	$0 \cdot 56\pi$
DMDMDM	$0 \cdot 35\pi$

Another method of sharpening the reflectance peaks is that using a combination of several filters of different orders.   A filter with a first-order peak at a frequency ν_0 has maxima at ν_0, $2\nu_0$, $3\nu_0$, etc. One with a first-order at $\nu_0/2$ has maxima at $\nu_0/2$, ν_0, $3\nu_0/2$, $2\nu_0$, etc. and one with a first-order maximum at $\nu_0/4$ possesses peaks at $\nu_0/4$, $\nu_0/2$, $3\nu_0/4$, $\nu_0, \ldots$, $2\nu_0, \ldots$ A combination of the three filters results in a system with very narrow reflectance peaks at ν_0, $2\nu_0$, $3\nu_0$, etc.   A reflectance band comparable in width with one of the maxima of the highest order filter is obtained, the lower order filters serving to suppress the side maxima.   Use has been made of filters of this kind in the infra-red region of the spectrum.   A typical three-filter combination made by Billings[9] using calcium fluoride and silver sulphide has the characteristic shown in *Figure 7.11*.

The multiple LH quarter-wave filter made entirely of dielectric layers, described in the last section, enables higher peak reflectances to be obtained than with the dielectric + metal system described above, in which the peak reflectance is only slightly higher than that of the opaque metal layer. The all-dielectric filter would now normally be used; the fact that a larger number of layers is required to give the same reflectance as the metal + dielectric system presents no difficulty when once the monitoring apparatus is assembled.

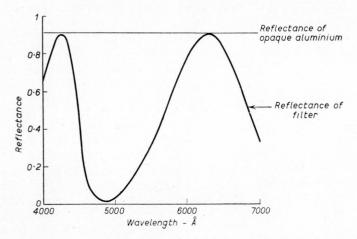

Figure 7.10   Experimental reflectance curve for system Al + MgF₂ + transparent Al

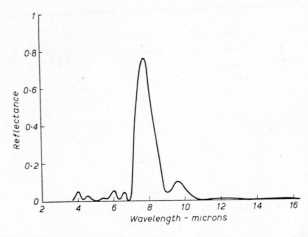

Figure 7.11   Experimental curve of reflectance of system of three CaF₂ + Ag₂S reflection filters

223

(c) *Cold mirrors*

An ingenious application combining the high reflectance obtainable by multiple LH $\lambda/4$ layers with the high infra-red transparency of semi-conducting materials has resulted in reflecting systems having a high reflectance in the visible spectrum and high transmittance in the near infra-red. The value of such a device—e.g. in a film projector, in which the power of the illuminating source is limited by the need to avoid undue heating of the film—is at once obvious. Turner [2] has made cold mirrors of this type, using a layer of germanium (of transmittance 17 per cent at $\lambda = 4400$ Å) covered by a LHLH system to give a reflectance peak at 5000 Å. A reflectance of at least 70 per cent is obtained between $\lambda = 4000$ Å and $\lambda = 6200$ Å, with a maximum of 90 per cent at $\lambda = 5000$ Å. The reflectance falls steeply to a few per cent at $\lambda = 7000$ Å. The average transmittance in the region $0\cdot8$–$2\cdot7\,\mu$ is about 80 per cent. In experiments comparing the performance of a silvered mirror with a cold mirror in a carbon-arc projector it was found that the cold mirror enabled the power of the lamp to be increased by 50 per cent for the same heating at the film gate.

7.4. ALL-DIELECTRIC HIGH-REFLECTING FILMS IN INTERFEROMETRY

(a) *The Fabry-Pérot interferometer*

Mention has already been made of the limitation in attainable resolution in interferometers in which silver reflecting films are used. The improvement to be expected in the performance of the Fabry-Pérot interferometer employing all-dielectric reflecting films in place of the conventional silver layers is considerable. The intensity distribution in the Fabry-Pérot system is given by

$$I = \frac{I_0}{1 + F \sin^2 \delta/2} \qquad \text{.... } 7(6)$$

where $I_0 =$ incident intensity and $F = 4\boldsymbol{R}/(1 - \boldsymbol{R})^2$, $\boldsymbol{R}$ being the reflectance of the film. $\delta\,(= \frac{2\pi}{\lambda} \cdot 2nt \cos \varphi)$ is the phase difference between successively reflected beams of wavelength λ traversing the interferometer at an angle of incidence φ. nt is the optical path between the reflecting surfaces, including any phase changes at these surfaces. In the absence of absorption, the intensity at the maximum in the rings is equal to the incident intensity. If the transmittance of the films is T and the absorptance A then the maximum intensity is given by

$$I_m = \frac{I_0}{(1 + A/T)^2} \qquad \text{.... } 7(7)$$

For high resolution, the reflectance R must be as high as possible, consistent with a usable amount of transmitted light. Typical values for a silver film of reflectance $R = 0.95$ are $T = 0.01$ and $A = 0.04$. From equation 7(7) it is seen that the intensity at the maximum of the fringes is only 0.04 of the incident intensity. We may contrast this with the value obtained by Ring and Wilcock[6] for seven-layer all-dielectric reflecting surfaces where $R = 0.983$, $T = 0.012$ and $A = 0.005$. In this case the transmitted intensity is 50 per cent of the incident intensity. The higher value of R moreover results in a smaller fringe width and hence in improved resolution.

The resolving power of the Fabry-Pérot etalon is conveniently considered in terms of the number of equivalent reflections, N, defined as the number of steps in a reflecting echelon which would give the same resolving power and order of interference. For new and aged silver films at $\lambda = 4200\,\text{Å}$, of such thickness that a transmittance of 0.45 at the peak is obtained, H. KUHN and B. A. WILSON[10] find N to be 16 and 11 respectively. For seven-layer dielectric films of zinc sulphide and cryolite, Ring and Wilcock obtain a value of about 250. The resolution suggested by this figure is not attainable in practice, the limit being set by the lack of flatness of the interferometer plates. However, whereas silver films do not allow full advantage to be taken of the best quality flats which can be made, the all-dielectric system is more than adequate.

The disadvantage of the dielectric system is that the high reflectance is attained only at specific wavelengths; if it is desired to work at several wavelengths, then separate interferometers for each wavelength are needed.

(b) *Low-order multiple-beam interferometry*

All-dielectric stacks have been used in place of silver films in systems for examining surface topography using multiple-beam Fizeau fringes (see Section 5.5). They are particularly valuable for the production of high-contrast Fizeau reflection fringes. The high absorption/transmission ratio for the thick silver films which are needed to ensure a narrow fringe width results in a low contrast between the fringes and the (bright) background. In order that reasonable visibility is attained, the reflectance of the silver needs to be kept to about 90 per cent with resultant loss of fringe sharpness. The lower absorption in multilayers enables reflectivities well above 90 per cent to be used.   J. A. BELK, S. TOLANSKY and D. TURNBULL[11] have used seven-layer ZnS + cryolite stacks having a reflectance of 95 per cent for the observation of fringes from a mica surface.   The high definition obtained is quite reproducible with such multilayers. A further advantage of the all-dielectric stack is that lower reflectances may be obtained with little absorption.   When viewing fringes between a surface and a reflecting flat, the reflectance of the flat should be approximately the same as that of the surface.   Too large a difference results in poor contrast in the fringe pattern. Thin silver layers generally show a very high A/T ratio whereas for all-dielectric layers, the absorption is low; high contrast is obtained, even with reflectances of only ∼40 per cent.

Belk *et al.* make interesting use of the selective reflectance of the all-dielectric stack by first illuminating a Fizeau fringe system with light for which the stack has a high reflectance.   A Fizeau fringe pattern results.   The wavelength of the light is then changed to that for which the reflectance of the stack is low.   The stack then transmits most of the incident light and a normal image of the surface is obtained.   This procedure is not possible with a silvered surface: in fact, the process of silvering may well mask surface features of interest.

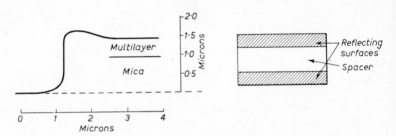

Figure 7.12   Multilayer contour in the immediate neighbourhood of a step on a mica surface

Figure 7.13   The interference filter

It is of importance to know whether multilayer stacks contour the surface on which they are deposited.   Only if they do so to a high degree of accuracy are they to be relied upon for interferometric topographical work on a fine scale.   By measuring the steps on cleaved mica and on crystals showing growth spirals, both with silver and all-dielectric reflecting surfaces, Belk *et al.* obtain the same values for the step heights to within the experimental error of about 10 Å. Experiments of S. TOLANSKY[12] have shown that the contouring by silver films is exact to within experimental error. A slight 'edge effect' is observed with multilayers in the neighbourhood of a sharp step, showing a slight tendency to pile up at the top of a step, (*Figure 7.12*). The irregularity is confined to the very near vicinity of the step and does not interfere with the measurement of film thickness by this method.

7.5. INTERFERENCE FILTERS

The name 'interference filter' is somewhat lacking in explicitness; there are many filters whose operations are governed by interference phenomena. This term has, however, come to refer to the Fabry-Pérot type of filter consisting of two highly reflecting surfaces separated by a spacer (*Figure 7.13*). When white light is passed through such a system and subsequently dispersed, a banded spectrum is seen. If the reflecting surfaces have a high reflectance, the transmission-bands in the spectrum are very narrow. The thickness of the spacer layer determines the separation of the transmission-bands. It is possible to arrange that only one such band occurs in the visible spectrum.

If t is the *optical* thickness of the filter, comprising the optical path in the spacer plus the path equivalent of the phase changes at the dielectric/reflecting surfaces, then transmission-bands occur at wavelengths λ such that

$$2t \cos \varphi = m\lambda \qquad \text{.... } 7(8)$$

where φ is the angle of incidence in the spacer film and m is an integer. Thus the system at normal incidence possesses transmission-peaks at wavelengths $2t$, t, $2t/3$, $t/2$, etc. If t is equal to, say, 5460 Å, then transmission-peaks are centred at 10,920 Å, 5460 Å, 3640 Å, 2730 Å. etc. The only peak in the visible is at 5460 Å so that such a filter would serve to isolate the mercury green line. The width W of the transmission-band at wavelength λ is given by

$$W = \frac{1 - R}{\pi m R^{\frac{1}{2}}} \cdot \lambda \qquad \text{... } 7(9)$$

where m is the order of interference. For the filter referred to above using silver films with $R = 0.90$, the bandwidth at $\lambda = 5460$ Å for which $m = 2$ is 84 Å. The filter will therefore effectively remove the adjacent yellow lines. Assuming an absorption of 0·04, the transmittance at the peak is 0·36. Although much higher values of T are attainable with this type of filter using all-dielectric reflecting layers, the filter performance when silver films are used compares very favourably with that obtainable with dye filters. The interference filter possesses the further advantage over the dye filter that its transmission peak may be placed anywhere at will (subject to the availability of non-absorbing materials for the layers).

We see from equation 7(8) that the bandwidth varies inversely as the order, suggesting the use of high-order filters when narrow bandwidths are required. The higher the order of interference at a given wavelength, the closer are the adjacent transmission-bands: this may be a disadvantage. If, however, it is required to select by means of the filter one spectral line from a not too crowded spectrum, it may be possible to choose the order of interference so that other transmission bands miss the unwanted spectral lines. Advantage may then be taken of the low bandwidth of the higher orders. It may then be possible to use slightly thinner silvering and thus to obtain a higher transmittance.

A limited measure of control over the position of the pass-band of an interference filter is possible by variation of the angle of incidence. Tilting from normal incidence moves the pass-bands in the direction of shorter wavelengths. For angles of incidence up to about 10° there is little change in the shape of the transmission-band. Beyond this angle, a broadening occurs and then a splitting into two separate bands, arising from the difference in the phase changes on reflection at the silver surface for components of the incident light polarized in or perpendicular to the plane of incidence. If plane polarized light is used with the filter, then large angles of tilt may be used. One half of the intensity is lost in the polarizer, however.

The films are deposited by thermal evaporation: details of monitoring are given in Section 7.9. Since such films are known to contour the substrate surface, the filter may be deposited on any reasonable piece of glass; a high quality optical flat is not required.

All-dielectric interference filters

Although some improvement in performance results from combining two or more silver + dielectric interference filters, e.g. by depositing five-layer systems of the type Ag-MgF$_2$-Ag-MgF$_2$-Ag (C. Dufour[13]) the inherent absorption of the metal reflecting film sets a severe limit to the narrowness of the band obtainable with reasonable light transmission. All-dielectric stacks in place of the silver enable vastly improved performance to be attained. Using the arrangement shown in *Figure 7.14*, Ring and Wilcock[6] obtain a bandwidth of 22 Å at $\lambda = 4600$ Å, with a peak transmission of 70 per cent. Increase of the number of LH pairs in the reflecting stacks produces a steady decrease in bandwidth until the stage is reached where the want of homogeneity in the spacer layer causes a variation in the position of the transmission-peak over the surface of the filter.

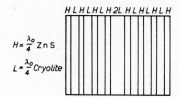

$H = \frac{\lambda_o}{4} Zn S$

$L = \frac{\lambda_o}{4} Cryolite$

Figure 7.14 All-dielectric interference filter

Thus when the transmission-band of a 39-layer filter is examined under high dispersion, an irregular line is obtained showing a variation in the wavelength of the transmission-band greater than the filter bandwidth. The variation observed is of the order 3-4 Å so that this would represent the practical lower limit of bandwidth attainable with filters employing evaporated layers. The super-narrow bandwidth filter provides a method for studying the optical homogeneity of the spacer layer. A similar, but rather more convenient method of doing this is afforded by the frustrated total reflection filter (Section 7.6).

Interference filters in the infra-red

The wealth of really high-index transparent materials in the infra-red makes this a highly suitable region for interference filters. Cryolite and magnesium fluoride remain transparent to wavelengths beyond 10 μ and are useful as low-index materials. Germanium is transparent beyond 1 μ and has a refractive index of 4·0. Beyond 4 μ, tellurium, with the impressive index of 5·4 enables high-reflecting stacks to be made with but few layers. *Table 7.5* shows the theoretical performance of various reflecting multilayers of tellurium and cryolite at a wavelength of 4 microns, at which neither material shows appreciable absorption.

The last two columns of the table illustrate the bandwidths and peak transmittances of interference filters made with such layers. The figures quoted for the bandwidths are the theoretical ones, which are generally approached fairly closely in practical filters. The peak transmissions are estimated from figures obtained from experimental filters which enable an estimate to be made of the loss of light by scattering in the films.

The main difficulty which is met in making multilayer stacks in the infra-red is that of depositing the rather thick layers required. Some materials tend to scatter rather badly in large thicknesses. Thermally evaporated films are generally in a state of strain which is compressive for some materials and tensile for others. Some degree of stress compensation is attainable by the judicious choice of

Table 7.5.   Te and Cryolite Multilayers on Quartz

System	Peak reflectance System in air at 4 μ	Interference filters. $\lambda_0 = 4$ μ		
		System	Band-width (μ)	Peak transmission (estimated)
Te . . .	0·82	QHLLH	0·29	0·90
Te + Crl + Te . .	0·98	QHLHLLHLH	0·025	0·84
Te + Crl + Te + Crl + Te	0·99₉	QHLHLHLLHLHLH	0·0014	0·78

materials and by the use of the correct thicknesses, although the optical characteristics required tend to tie down the refractive indices and the thicknesses.   There is some measure of freedom in the design, however, insofar as films of thickness $\lambda/4$, $3\lambda/4$, $5\lambda/4$ . . . behave similarly in a multilayer stack at the wavelength λ.   If control of the properties of the system over only a short wavelength range is required, then the latitude offered by such choices of thickness is invaluable in achieving a measure of stress-compensation.

Influence of inaccuracies in the layer thicknesses

The very high performance of which the high-reflecting, all-dielectric multilayers have been shown to be capable is of limited usefulness if the accuracy required in the layer thicknesses is inordinately difficult to attain experimentally.   In fact, the tolerances on the layer thicknesses are surprisingly high.   C. DUFOUR and A. HERPIN[14] consider the effect of an error in one layer on the form of the reflectance curve and show that the reflectance in the neighbourhood of the peak is but slightly altered for a thickness error of 10 per cent in a high-index layer.   The labour of calculating the effect of small errors is lessened by an approximation method (O. S. HEAVENS[15]) which shows that the change in reflectance at the peak vanishes, to a first order, for a first order error in one of the layer thicknesses.

It is possible to control the thickness of the depositing films to within ± 5 per cent with quite simple auxiliary apparatus (Section 7.9) so that the construction of high-reflecting stacks is a matter of no difficulty. The interference filter presents a more severe problem since the position of the transmission-band is determined by the thickness of the spacer layer. The very narrow bandwidths which are so readily obtained using multilayer stacks impose the need for high accuracy in the thickness of the spacer layer. A 1 per cent error in the spacer thickness for a filter centred at $\lambda = 5000$ Å results in a transmission peak wide of the mark by 50 Å. If the bandwidth of the filter is only 20 Å, the transmittance of the filter at $\lambda = 5000$ Å would be very low. Control of the thickness of the spacer layer requires elaborate and sophisticated techniques which are described in Section 7.9 (c) below.

7.6. THE FRUSTRATED TOTAL REFLECTION FILTER

The bandwidth of the all-dielectric multilayer interference filter described in Section 7.5 may be made extremely narrow by the use of suitably large reflecting stacks. The 15-layer filter of Ring and Wilcock has a bandwidth of 22 Å. The absorption of ZnS and cryolite in the visible region of the spectrum is not sufficiently high to account for the loss of 30 per cent at the transmission-peak and much of this loss probably arises from light scattering in the layers. The granular structure of thermally-evaporated films inevitably gives rise to some scattering and this effect is found to become worse with increasing number of layers.

An ingenious alternative method of obtaining the high reflectance required for the interference filter with no absorption and a minimum of trouble from scattering is that using frustrated total reflection, devised by Turner[2]. Under conditions of total reflection at a surface separating semi-infinite regions, the reflectance is unity (for a beam of infinite cross-section). The disturbance in the second medium decays exponentially with distance normal to the boundary and the value of the Poynting vector in the second medium is zero. When total reflection occurs at a surface covered by a low-index film, then some light is transmitted by the film and the remainder reflected. The proportion of light transmitted is governed by the film thickness and depends on the state of polarization of the light.

Any value of reflectance may be obtained by adjustment of the thickness of the frustrating layer. Since no absorbing materials are used, there is no light loss from absorption: since only three films are required (frustrating layer + spacer + frustrating layer) there is much less light loss from scattering by the films of the filter. As the change

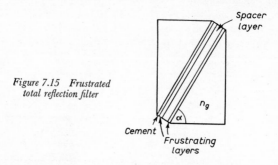

Figure 7.15   Frustrated total reflection filter

in phase on reflection at the spacer/frustrating layer interface depends on the state of polarization of the incident light, the filter exhibits two transmission bands, when used with unpolarized light, with a theoretical transmittance of 50 per cent in each band. The dispersion of such a filter taking account of the dispersion of the film material, is given by H. D. POLSTER[16].

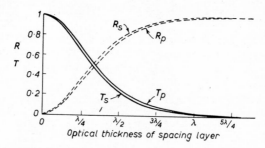

Figure 7.16   Variation of reflectance and transmittance of totally reflecting surface covered by frustrating film

The usual form of the frustrated total reflection filter is shown in *Figure 7.15*. The indices of the glass and frustrating layers are chosen so that the critical angle is less than the angle a of the prism. Light may then be directed normally on the prism base. For a frustrating layer of MgF_2 ($n = 1.38$), a 60° prism of dense flint glass of index 1·72 is suitable, the critical angle for the combination being about 53°. The first frustrating layer is evaporated on to the hypotenuse face of one of the prisms, followed by a high-index spacer layer. Since the light passes through the spacer layer at non-normal incidence, the spacer thickness t_s, required to give a maximum for wavelength λ is given by

$$2n_s t_s \cos \varphi_s = m\lambda \qquad \dots 7(10)$$

where $n_s \sin \varphi_s = n_g \sin \varphi_g$. φ_s, φ_g are the angles of incidence in the spacer and glass respectively. A second frustrating layer is then evaporated on to the spacer and the second prism is cemented to the filmed surface of the first by a cement of index equal to that of the glass.

Figure 7.16 shows the variation of **R** and **T** with optical thickness of a frustrating layer of index 1·38 on a prism of index 1·72, the angle of incidence being 60° (Turner[2]). Experimental filters made with flint prisms and MgF_2 frustrating layers and using ZnS as the spacer layer show that, for optical thicknesses of spacer layer up to $\sim 1·1$ wavelengths, the filter characteristics obtained are fairly close to those expected from theory. Peak transmissions of ~ 93 per cent of the expected values are attained with bandwidths down to about 60 Å. Increase of the frustrating layer thickness to 1·3 λ resulted in a bandwidth of 30 Å instead of the theoretically expected value of 12 Å. The peak transmission was appreciably below the expected value. This failure is due to the lack of uniformity of the optical thickness of the spacer layer, which arises in part from the roughness of the surface of the frustrating layer. As with the interference filter, use cannot at this stage be made of the narrow bandwidths potentially available with this type of filter on account of the lack of homogeneity of the evaporated films employed.

Use of the frustrated total reflection filter to examine film homogeneity

The variation in wavelength of the transmission-peak of the filter is related to variation in the optical thickness t of the spacer film (+phase changes) by

$$\frac{d\lambda}{\lambda} = \frac{dt}{t} \qquad \dots 7(11)$$

so that variation in thickness may in principle be deduced from the variation in the wavelength of the transmission-peak over the filter surface. Direct observation with a spectrometer yields a result limited by the dispersion of the instrument. A more sensitive arrangement, used by A. E. GEE and H. D. POLSTER[17], is shown in *Figure 7.17*. The spectroscope is illuminated with a line source corresponding to the transmission-peak of the filter. If the spacer thickness is exactly uniform, then high transmittance is obtained over the whole area of the filter, which is viewed by removing the telescope eyepiece. Any variation in thickness of the spacer, with consequent shift of the transmission maximum, is manifested by a falling-off in the transmitted intensity. If the bandwidth of the filter is, say, 5 Å, then practically no light passes through regions

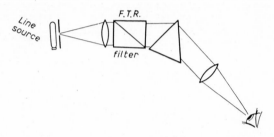

Figure 7.17   Arrangement for studying homogeneity of the spacer layer of the frustrated total reflection filter

whose thicknesses differ from that corresponding to the peak transmittance by more than about 5 Å. The intensity distribution seen in the field is thus a highly sensitive contour map of the optical thickness of the spacer—sensitive, that is, over the range $\pm H$, where H is the filter half width. A convenient way of interpreting the intensity diagrams obtained is to examine the change in the contour pattern as the filter is tilted. A sensitivity of 0·5–1 Å in film thickness is attained by this method.

These experiments show clearly the reason why the theoretically expected narrow bandwidths are not attained with the frustrated total reflection filter. Variations in optical thickness of up to 10–15 Å are observed. The use of the filter in this way affords a powerful method for studying the uniformity of evaporated films. Gee and Polster examine the effect of rotating the target surface during evaporation and of varying the rate of deposition. Decidedly better uniformity is obtained when the rate of deposition is kept to about 10 Å/sec than when higher rates (100 Å/sec) are used.

The birefringent frustrated total reflection filter

Work on the frustrated total reflection filter offers examples of boundless ingenuity. The use of frustrating layers as non-absorbing high-reflecting films may be regarded perhaps as the first stroke of genius. The second is provided by B. H. BILLINGS[18] in the method used to persuade the filter to show a single transmission-band instead of two. The two bands arise apparently unavoidably from the difference in phase changes on reflection for the two components of polarization. This phase difference is compensated, in the Billings modification, by using a birefringent spacer layer. It is arranged that the difference in optical path through the birefringent layer is equal and opposite to that arising from the reflection phase changes. There are several organic materials which may be evaporated to form birefringent films; they generally consist of planar molecules which tend to lie flat on the substrate, so producing a film with its optic axis normal to the substrate surface. The refractive indices of these materials, of which uric acid is typical, are generally high, so that they are conveniently used as the spacer layer of the filter. The two transmission bands of the filter are brought into coincidence by adjustment of the angle of incidence. A slight disadvantage of the system of using the birefringent material as the spacer layer is that a different angle of incidence is required for different interference orders. If the frustrating layers are made birefringent, instead of the spacer layer, then this difficulty is overcome. Unfortunately the known materials suitable for making birefringent layers have high refractive indices.

7.7. COMPARISON OF MULTILAYER FILTERS WITH OTHER NARROW PASS-BAND FILTERS

A glance at the transmission characteristics of dye filters suffices to indicate that only in rare cases can reasonably narrow transmission bands be obtained by these filters. For transmittances of about 50 per cent, the order of magnitude of bandwidth is measured in hundreds of ångströms. When used in conjunction with a line source in which there are few, widely spaced lines, combinations of dye filters may enable a practically monochromatic source to be obtained with but little light loss. Where the source lines are close together, separation with dye filters can be only imperfectly attained and then only with considerable energy loss.

In the astronomical field, the need has arisen for a narrow-band filter for studying the solar corona. Examination of the corona by means of the light emitted by the various elements present enables the distribution of those elements to be determined. The distribution of hydrogen may be measured by photographing the solar corona in the light of the $H\alpha(\lambda = 6563$ Å$)$ line. In view of the large amount of light present of neighbouring wavelengths, a filter of extremely narrow bandwidth—of the order 1–2 Å—is required if reasonable discrimination is to be attained.

The polarization filter devised by B. LYOT[19] may be made with a bandwidth of 1 Å. Use is made in this system of a set of birefringent plates separated by polarizers. If n_e and n_o are the refractive indices of a plate of thickness d, for light of wavelength λ traversing normally, then the intensity of the light transmitted by the plate when sandwiched by parallel polarizers is given by

$$I = I_0 \cos^2 \left\{ \frac{\pi d(n_e - n_o)}{\lambda} \right\} \qquad \text{.... } 7(12)$$

For a given plate, maxima in the transmitted light occur at wavelengths λ_0, $\lambda_0/2$, $\lambda_0/3$, . . . etc. where $\lambda_0 = d(n_e - n_o)$. The wavelength separation between transmission maxima increases with decreasing plate thickness. If several plates with thicknesses decreasing in arithmetic progression are combined, their individual transmission curves are as shown in *Figure 7.18*, together with the curve for the combination of all the plates. The bandwidth of the system is governed by that of the bands from the thickest plate. This sets a practical limit to the narrowness of the band obtainable. For a bandwidth of 1 Å at the $H\alpha$-line, and using quartz as the birefringent material, the required thickness for the largest plate is 23·84 cm. Thus, for a filter of reasonably large aperture, a pretty big block is needed; it is doubtful whether a block of quartz of good enough optical quality of sufficient size could be found. Use has been made of more strongly birefringent materials, for which much smaller thicknesses are needed. B. H. BILLINGS, *et al.*[20] have used ammonium dihydrogen phosphate in a filter which achieves the desired 1 Å pass-band at $\lambda = 6563$ Å. Details are given of the construction of the filter and show that the required tolerances, on the thicknesses of plates, are attainable by the use of mica corrector plates. Billings's final filter comprises no less than 39 elements and is mounted in a temperature-controlled housing in order that the $\pm 0·08°$ C needed to hold the pass-band in the right place can be obtained.

From the discussion of multilayer filters in the previous sections, we see that at the present stage of development, bandwidths as small as 1 Å are unattainable, on account of the difficulty in forming layers of sufficient optical homogeneity. The work of Gee and Polster suggests that some improvement is to be expected over the 20–30 Å bandwidths now attainable with reasonably high transmission by an investigation into the effect of evaporation conditions. From the nature of the structure of thermally evaporated layers, it is clear that there will always be some degree of inhomogeneity; we may guess that the variation in optical thickness from this cause may be brought down to a small number of ångström units. The light loss from

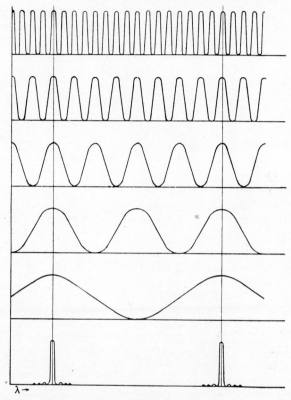

Figure 7.18   Transmission curves of birefringent plates separated by polarizers

scattering, which afflicts particularly the all-dielectric interference filter, may also be improved upon in this way.

The simplicity of the multilayer filter, either of the Fabry-Pérot type or the Turner frustrated total reflection type, commends this form of filter in cases where bandwidths of the order 20 Å can be tolerated. Although the apparatus for accurate monitoring is a little complicated, its assembly is a 'once-for-all' task and filters may then subsequently be made rapidly and easily. The labour involved in making the polarization filter consists largely in accurate polishing of the birefringent crystals and preparation and adjustment of corrector plates; this must needs be done separately for each filter made.

No details are given of the transmittance of the polarization filter. Since, in the case of the Billings filter, there are 39 elements, the reflection loss at the 78 surfaces would normally amount to over 90 per cent. Single anti-reflecting films, applied to each of the 78 surfaces may be expected to reduce the reflection loss to about 30 per cent. Double-layer anti-reflecting films would further reduce this loss, although the construction of the filter is by now becoming a somewhat formidable proposition. The multilayer interference filters attain a favourably high transmission without complication by reflection loss.

Scattering loss in the frustrated total reflection filter is less than that of an all-dielectric multilayer Fabry-Pérot type filter on account of the smaller number of layers used. Large aperture filters of the frustrated total reflection type are likely to be expensive in view of the large prisms which are needed. Until the technique is developed of depositing films of sufficient homogeneity to enable use to be made of the extremely narrow bandwidth which is so simply obtainable with the frustrated total reflection filter, the all-dielectric multilayer stack would appear to be the easier approach to a moderately narrow-band filter. But for bandwidths in the 1 Å order, the polarization filter is needed.

7.8. The Use of Thin Films as Polarizers

The need for an efficient polarizer of large aperture may be met in a quite simple fashion by the use of thin transparent layers. Apertures vastly greater than the inevitably limited ones attainable with polarizing prisms of the Nicol type are readily attained. A high degree of polarization is obtainable with few elements and with very high efficiency. The principle of using a pile of plates at the Brewster angle for producing a reflected beam with a high degree of polarization is long established. Its main disadvantages lie in the large number of plates required in order to obtain an appreciable amount of energy in the beam.

By suitable choice of the thickness and refractive index of a film and of the angle of incidence of light on the film, it is possible to arrange suppression of one of the polarized components while at the same time enhancing the reflectance of the perpendicular component.

238

The condition for zero reflectance of a film of thickness d_1 and refractive index n_1 is given, from equation 4(44) by

$$r_{1p,s} + r_{2p,s}e^{-2i\delta_1} = 0 \qquad \text{.... } 7(13)$$

where r_1, r_2 are the Fresnel reflection coefficients for the p or s components of polarization and $\delta_1 = 2\pi n_1 d_1 \cos \varphi_1 / \lambda$, φ_1 being the angle of incidence of the light in the film. In order to satisfy equation 7(13) we require that

$$r_1 = r_2 \qquad \text{.... } 7(14)$$

$$2\delta_1 = (2m+1)\lambda \qquad \text{.... } 7(15)$$

Since we are concerned with non-normal incidence, the values of r_1, r_2 depend both on the refractive indices and on the angle of incidence of the light. We may thus choose the angle of incidence to suit any particular value of n_1. The thickness of film of index n_1 required for suppressing reflection is, from 7(15),

$$d_1 = \frac{(2m+1)\lambda}{4n_1 \cos \varphi_1} \qquad \text{.... } 7(16)$$

Consider the component polarized with its electric vector in the plane of incidence. From equations 7(14) and 4(19) we may deduce the relation between φ_0, the angle of incidence in the first medium n_0 (usually air) and the film index n_1.

We obtain

$$n_0^2(n_1^8 - n_0^4 n_2^4) \sin^4 \varphi_0 + n_1^2 \{2n_0^4 n_2^4 - n_1^6(n_0^2 + n_2^2)\} \sin^2 \varphi_0 + n_1^4 n_2^2(n_1^4 - n_0^2 n_2^2) = 0 \qquad \text{.... } 7(17)$$

The roots of this equation are real for $n_1^2 > n_0 n_2$. Values of φ_0 as a function of n_1 are plotted in *Figure 7.19* for substrates of index 1·5 and 1·7.

Thus for a given substrate and film index n_1 we may calculate the thickness of film d_1 required for perfect suppression of $\boldsymbol{R}_p$ and the angle of incidence φ_0 at which the zero reflectance occurs. The choice of n_1 remains arbitrary in that zero of $\boldsymbol{R}_p$ may always be attained by adjustment of d_1 and φ_0. If the film is to serve usefully as a polarizer, then the perpendicular component $\boldsymbol{R}_s$ must needs have as large a value as possible. Since the value of $\boldsymbol{R}_s$ increases with increasing n_1, the highest practicable value of n_1 is chosen. $\boldsymbol{R}_s$ is given by

$$R_s = \left(\frac{n_1^2 \cos^2 \varphi_1 - n_0 n_2 \cos \varphi_0 \cos \varphi_2}{n_1^2 \cos^2 \varphi_1 + n_0 n_2 \cos \varphi_0 \cos \varphi_2}\right)^2 \qquad \dots 7(18)$$

when the condition for zero R_p obtains.

F. ABELÈS[21] has studied the properties of titanium-dioxide films as polarizers, deposited on glass of index 1·53. The thickness was chosen for minimizing R_p at $\lambda = 5500$ Å. The degree of polarization of the transmitted light is defined by

$$P = \frac{T_p - T_s}{T_p + T_s} = \frac{1 - T_s}{1 + T_s} \qquad \dots 7(19)$$

since $T_p = 1$ by virtue of the fact that $R_p = 0$ and the film is non-absorbing. For a glass plate with a TiO_2 film on each face, the value of P obtained was 0·92. Two such plates, placed parallel to one another, yielded transmitted light for which $P = 0·997$. The superiority over the pile-of-plates is brought out by the fact that twelve such plates are required in order to obtain a degree of polarization of 0·96.

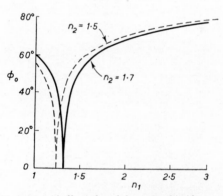

Figure 7.19 Variation with film index of the angle of incidence at which zero reflectance occurs for substrates of index 1·5 and 1·7

If such a system is to be used for white light, it is important that the performance of the polarizer shall not deteriorate rapidly with change of wavelength. For the system described above, using TiO_2 layers, the reflectances at various wavelengths are shown in *Table 7.6.* The dependence on wavelength of the degree of polarization P is smaller for a pair of plates, each coated with TiO_2 on both faces. From 0·997 at the centre of the visible spectrum, P falls only to 0·993 at the ends of the spectrum.

Table 7.6

λ (Å)	R_p	R_s	$P*$
4000	0·03	0·73	0·86
5500	0	0·787	0·92
8000	0·02	0·75	0·88

* The value of *P* is for a glass plate coated on each face with a TiO_2 layer.

Polarizing beam splitters

The disadvantages of metal films as partial reflectors for dividing a beam of light into equal parts are well known. At the thicknesses which yield equality of the reflected and transmitted beams, the film absorption takes an unhealthily large share of the incident energy. As with the interference filter, this difficulty can be very largely overcome by the use of dielectric layers. By suitable choice of the refractive index and dispersion of the glass supports, a beam-splitting system may be easily made from which the two emergent components of the light beam are plane polarized to a reasonably high (0·98) degree of polarization.

The arrangement devised by M. BANNING [22] is shown in *Figure 7.20*. By suitable choice of the refractive index of the prism material, it is arranged that light strikes the glass/zinc sulphide and sulphide/cryolite interfaces at the Brewster angle. The condition that this should occur is seen, from Brewster's and Snell's laws, to be

$$n_3^2 = 2n_1^2 n_2^2/(n_1^2 + n_2^2) \qquad \text{.... } 7(20)$$

If the beam splitter is to be used with white light, there is a further consideration to be met. In order that the Brewster condition is fulfilled at all wavelengths, the dispersion of the glass must be correctly related to that of the film materials. For cryolite, the dispersion is negligibly small in the visible. For zinc sulphide, the ν-number[†] is 17. Banning shows that, with $n_1 = 2\cdot30$ and $n_2 = 1\cdot25$ (corresponding to the *couche lacunaire* obtained by evaporating cryolite at a pressure of 10^{-3} mm Hg), the glass should have a refractive index of 1·55 and a ν-value of 48·5. These are easily realizable.

The beam splitter shown in *Figure 7.20* was found to give reflected light with a degree of polarization *P* of $> 0\cdot99$. For the transmitted

$$\dagger \nu = (n_{blue} - n_{red})/(n_{yellow} - 1)$$

241

light, P ranged from 0·997 in the green to a minimum of 0·945 in the blue. Over a range of incidence of $\pm 5°$ from normal, the average value of P for white light was 0·98.

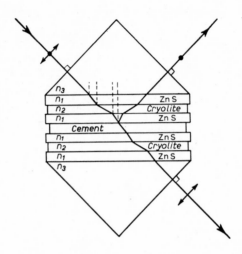

Figure 7.20 Banning's polarizing beam splitter

7.9. METHODS OF MONITORING FILM THICKNESS

The techniques which have been developed for enabling films of any desired thickness to be deposited fall into three main groups, depending on the accuracy required. The light reflected or transmitted by the film system is examined during deposition and the thickness related either to the hue observed or to the intensity of light of a particular wavelength. For dielectric films, the intensity of the reflected and transmitted light passes through maxima and minima as the thickness increases and these turning-points serve to indicate when the desired thickness is attained. Since the phase change at boundaries of non-absorbing layers is always either zero or π, no difficulty arises of unknown phase changes as is the case with absorbing materials. For monitoring heavily absorbing materials, the attenuation of a light beam traversing the film is sometimes used. A calibration is first made, in which the film thickness may be determined by Fizeau or Feco fringes (Section 5.5). Some care is needed since the transmittance of some metal films does not define the thickness uniquely (see e.g., *Figure 6.10*).

242

(a) *Monitoring by visual inspection (Banning[22])*

Although it is not capable of high precision, this method, which involves visual examination of light reflected by the film is sufficiently accurate for many purposes. It possesses the useful property of requiring no more than a lamp and a pair of eyes. Where several layers are to be deposited, it is convenient to monitor each layer separately on a fresh glass plate. Some guide is given below to the problem of dealing with films of ZnS and cryolite by this method although from the subjective nature of the process, the help

Table 7.7. *Appearance of ZnS and Cryolite Films of various Thicknesses*

Optical thickness ($\lambda = 5461$ Å)	Colour	
	ZnS	Cryolite
$\lambda/4$	Bluish-white	yellow
	White	magenta
	Greenish-yellow	blue
$\lambda/2$	Magenta	white
	Blue	yellow
$3\lambda/4$	Greenish-white	magenta
	Yellow	blue
λ	Magenta	greenish white

is necessarily limited. With practice, systems of films may be deposited possessing reflectances within 2–3 per cent of the theoretical values, except for high-reflectance systems where the reflectance exceeds 90 per cent. As mentioned in Section 7.5, quite large errors can be tolerated in the thicknesses of the layers in LH quarter-wave systems without seriously affecting the reflectance at the peak.

An alternative visual method of monitoring is to prepare in advance a number of films of various known thicknesses and to judge the thickness of the film being deposited by comparison of the light reflected from the film with that from the standard. These two methods are both of sufficient accuracy for the preparation of high-reflecting stacks except perhaps where the highest of performances is required.

(b) *Monitoring by intensity measurement (C. Dufour[23], D. T. Turnbull and J. A. Belk[24])*

A considerable improvement over the above method is that in which the intensity of the light reflected or transmitted by the film is measured using a photo-electric cell or other suitable detector. By suitable choice of the wavelength, the desired thickness can be made to correspond either to a maximum or a minimum. In principle, a method which entails stopping the evaporation when a meter-reading goes through a maximum or minimum is unsatisfactory. One never knows whether one has reached the turning-point until one has passed it. In practice, the error introduced by a slight

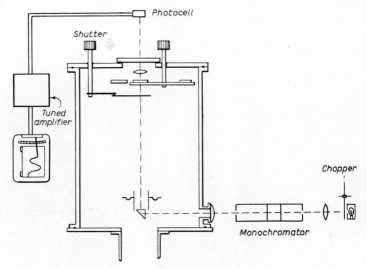

Figure 7.21   Arrangement of apparatus for monitoring photo-electrically

overshooting is for most purposes not serious. The difficulty is completely overcome by the method described in (c) below. When film stacks are required to give reflectance maxima or minima at a wavelength corresponding to a common spectral line, the light source may conveniently consist of a discharge lamp and suitable colour filter. For wavelengths which do not coincide with a readily separable spectrum line, a monochromator is required. The errors involved in tuning to a maximum or minimum are substantially reduced if a recording meter is used on the detector output. It is much easier to judge the position of the true turning value from a pen-record of the changing output than from a pointer-reading. A further advantage of this slightly more elaborate set-up is that, with a continuous drive on the wavelength control of the monochromator, the spectral characteristic of the finished multilayer may be readily obtained.

A schematic arrangement for monitoring by the above method is shown in *Figure 7.21*. The use of a chopped light-source and tuned amplifier avoids untoward interference from the light from the evaporation sources. The method can be easily applied to the ultra-violet and infra-red spectral regions as well as to the visible. The accuracy attainable is ample for all purposes except that of depositing the spacer layer in a narrow-band interference filter, or frustrated total reflection filter.

(c) *P. Giacomo and P. Jacquinot's method*[25]

The spacer layer of an interference filter presents a severe problem in monitoring. The progress in making all-dielectric high-reflecting stacks has enabled filters with transmission bandwidths of the order 20 Å to be made without great difficulty. This figure is, moreover, likely to be reduced as improvements in homogeneity of the layers are achieved. If such a filter is to be made to isolate a spectral line, then the centre of the transmission-band needs to be located with very high precision. It is doubtful whether the methods given above would enable the centre of the band to be placed reliably to within ± 100 Å. Some slight control of the position of the band can be effected by tilting the filter: this enables the band to be moved to shorter wavelengths only.

The problem has been dealt with in an elegant way by Giacomo and Jacquinot. The layer is illuminated by a beam in which the wavelength is varying over a small range. The beam falls on a detector from which the output is amplified. Since the transmittance of the film varies with wavelength, the amplifier output will in general contain a signal of frequency equal to the frequency of the wavelength scan (f). If the wavelength at the centre of the scanning range is one at which a maximum or minimum of transmission of the film occurs, then the output at frequency f will be zero. It is further possible to locate points of inflexion in the transmission vs. wavelength curve; for such points, the component of photo current of frequency $2f$ vanishes. Account must be taken of the variation with wavelength of the response of the detecting system used.

If the monitoring wavelength scans sinusoidally over a range $\pm a$ from the value λ_0, the wavelength at any instant is given by

$$\lambda = \lambda_0 + a \cos 2\pi f t \qquad \dots \ 7(21)$$

The photocell current i will be a function of the response of the detector system, A, which is generally wavelength-dependent and of the transmission curve $y = g(\lambda)$ of the film under examination. At time t, the output is given by

$$i = Ag(\lambda) = Ag(\lambda_0 + a \cos 2\pi ft) \qquad .... \; 7(22)$$

Neglecting for the moment the wavelength dependence of A, we may expand equation $7(22)$ as a power series in a and, noting that we make a small compared with λ_0, we retain only the first two terms. Then

$$i = A\left\{ g(\lambda_0) + a\left(\frac{dg}{d\lambda}\right)_0 \cos 2\pi ft + a^2\left(\frac{d^2g}{d\lambda^2}\right)_0 \cos^2 2\pi ft \right\} \quad .... \; 7(23)$$

The last term may be expressed in terms of $\cos 4\pi ft$ so that the output current is made up of (i) a constant component of magnitude $\simeq Ag(\lambda_0)$; (ii) a component of frequency f and of amplitude $Aa\left(\dfrac{dg}{d\lambda}\right)_0$ and (iii) a component of frequency $2f$ and of amplitude $\dfrac{Aa^2}{4}\left(\dfrac{d^2g}{d\lambda^2}\right)_0$. Vanishing of the f component thus indicates that $\left(\dfrac{dg}{d\lambda}\right)_0$ is zero—i.e. that the transmittance of the film, or film system, is a maximum or minimum. If the component of frequency $2f$ vanishes, then a point of inflexion is indicated. (Vanishing of both components simultaneously shows a point of inflexion with zero slope—a rarity in practice.)

The variation of A with wavelength is determined by the spectral distribution of the light source, the geometry and transmission of the optical system, the photocell response and the amplifier gain. These will generally contribute to give a fairly broad maximum in some part of the spectral range. Except when λ_0 is very near to this maximum, the effect of the variation of A with wavelength is to cause an error in the value of λ corresponding to the maximum or minimum. Thus if *Figure 7.22* (*a*) gives the transmittance, $g(\lambda)$, of the film and if (*b*) shows the variation of A, then the resultant signal vs. wavelength curve will be as in (*c*) and the wavelength of the maximum will be at λ_1 instead of at λ_0.

A simple way in which allowance may be made for the wavelength dependence of A is to arrange a razor-blade over the spectrum formed by the monochromator so that, over the small wavelength range $\pm a$ around λ_0, the variation of $A(\lambda)$ with wavelength is compensated by the variation in the energy output of the mono-chromator. Provided the variation of A with λ over the range $\pm a$ is not violent, sufficient compensation is obtained.

Experimental arrangements

The optical arrangement is as shown in *Figure 7.23*. The white light-source S, fed from a stabilized supply, is focused on the entrance slit of the monochromator. The film to be examined is placed in the beam before the slit. A spectrum is formed at C,

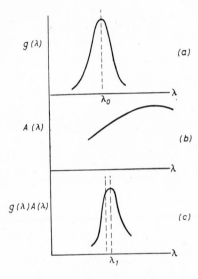

Figure 7.22

whereat the knife-edge for compensating the variation of A is situated. Light continues to the mirror M, attached to a spring which is caused to vibrate by the electro-magnet E, and thence is focused by the lens L on the photocell P. The angular vibration of M was about 6′, this serving to provide a wavelength scan of ± 100 Å at a frequency of 100 c/s. It will be noticed that the light passing the test-plate $g(\lambda)$ is not collimated. In practice, the convergence of the light falling on the test-plate is limited so as not to exceed 2–3°. The difference in the optical paths for the central and extreme rays is less than 0·15 per cent, corresponding to a difference in pass-bands at $\lambda = 5000$ Å of only ~ 7 Å. This has the effect of a slight broadening of the band and does not impair appreciably the accuracy of locating the centre of the band.

An electron-multiplier photocell is used, with a compensating circuit which ensures a constant over-all sensitivity. The output of the (two-stage) amplifier is fed to the Y-plates of a cathode-ray oscillograph, the X-plates being supplied from a 50 c/s source. Frequencies of f(100 c/s) and $2f$(200 c/s) are thus readily identified from the Lissajou figures obtained.

The centre of the pass-band of a narrow band interference filter

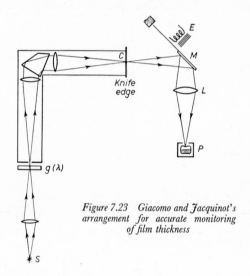

Figure 7.23   Giacomo and Jacquinot's arrangement for accurate monitoring of film thickness

may be located to within a few ångström units and the bandwidth (from the wavelength separation of the points of inflexion), to within a few per cent. By a simple iterative process, the position of the maximum of a pass-band may be located with high accuracy even when the bandwidth is large. Giacomo and Jacquinot obtain results consistent to within 1–2 Å for the transmission peak of a filter with a bandwidth of about 600 Å.

(d) *Greenland and Billington's method*

The method described in (c) above may be readily adapted for the deposition of films of any thickness by selecting as the centre of the scanning wavelength a value for which the optical thickness of the film yields a maximum or minimum in the transmitted (or reflected) light. It may be used for constructing any of the variety of multilayer systems described in Sections 7.2–7.6. A novel method by K. M. GREENLAND and C. BILLINGTON[26] enables the spacer layer of the Fabry-Pérot type interference filter to be monitored with reasonable accuracy and with no elaborate auxiliary apparatus. Although this was devised before the development of all-dielectric filters, the method may be used with such systems as well as with those employing silver layers. As mentioned above, the accuracy required of the quarter-wave layers in the high-reflecting stacks is not high. The visual method described in (a) above suffices for all but the most stringent requirements. The Greenland method enables interference filters to be made with the pass-bands located to within about 20 Å.

The position of the transmission-band of the silver + dielectric interference filter depends both on the optical thickness of the spacer layer and on the phase changes on reflection of the light at the silver/dielectric interfaces. Control of the position of the band is by the thickness of the dielectric layer. The bands cannot, however, be observed, at least at normal incidence, during the deposition of the dielectric since they are not there until the second silver layer is deposited; by which time it is a little late to do anything about the spacer layer thickness if this should prove incorrect. If, however, light is reflected from the silvered side of the filter during the deposition of the spacer, at such an angle of incidence that total reflection occurs at the dielectric/vacuum interface, then the system, incomplete at normal incidence, behaves as a reflection filter for angles of incidence beyond the critical. The dispersed reflected light contains sharp absorption bands which may readily be observed. Polarized light is used since, owing to the difference in phase changes for different states of polarization, the reflected light would with unpolarized light show two absorption bands. Each would, however, be superposed on the bright background of the other component, so that contrast would be poor. The wavelength at which the absorption-band occurs is found to be linearly related to that of the transmission-band at normal incidence, provided steps are taken to ensure that the silver layers are of the same thickness. It was found to be sufficient to adjust the silver film thickness by visual comparison of the light transmitted by the silver film with that of a standard.

7.10. MISCELLANEOUS APPLICATIONS OF THIN FILMS

By way of conclusion, we shall give brief mention of some applications of thin film properties which are mainly of a purely practical nature. Our main preoccupation thus far has been with films of uniform thickness. We shall see that results of considerable importance may be obtained in the optical field by the use of films of judiciously-chosen non-uniformity. It is probably quite safe to claim, in the case of many of these films, that no other known method would enable the often peculiar distributions of material to be obtained. In other cases, there is the possibility of replacing existing methods which are difficult and expensive by much simpler techniques. The applications considered are but a selection from a wide field and may serve to indicate the potentialities of thin film techniques in new problems.

(a) *Use of evaporated layers in producing aspherical surfaces*

Until recent years, the development of applied geometrical optics had been mainly limited to the use of spherical surfaces. The difficulties of grinding and polishing aspherical surfaces were such that advantage could not readily be taken of the improvements

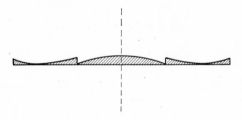

Figure 7.24   Profile of Fresnel mirror made by evaporation

which could be achieved by their use. Parabolic mirrors were made by figuring the nearest spherical surface; although not too difficult, the process is lengthy and expensive. More exotic profiles than the parabola, with which many of the defects of aberration could be reduced, are prohibitively difficult and expensive to make.

The method of thermal evaporation lends itself well to the problem of depositing odd-shaped distributions. J. STRONG[27] has used the method for parabolizing spherical mirrors. Since such mirrors are normally aluminized, the parabolizing deposit is made of aluminium. One of the limits to this process is set by the fact that evaporated aluminium films more than about 2 microns thick scatter light rather badly. L. G. SCHULZ[28] has shown how this difficulty may be overcome by the use of stepped deposits (*Figure 7.24*), forming a Fresnel mirror. Profiles of the type shown in *Figure 7.24* would be enormously troublesome to grind and polish.

250

They are simply produced by the evaporation method by the use of the rotating diaphragm shown in *Figure 7.25*.

An even greater contribution is made by evaporated films in the production of off-axis paraboloids, so useful in an optical set-up in which the wavelength used makes the use of refracting components difficult. This type of mirror can be made by conventional grinding and polishing by first making a complete paraboloid of the requisite size and cutting the off-axis mirrors from the complete mirror. For mirrors of large aperture, this procedure is very costly. The

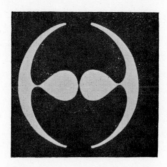

Figure 7.25

evaporation method enables the off-axis paraboloid to be made from a spherical blank of the required final size. Schulz has made off-axis mirrors of the Fresnel type by the evaporation method and finds their performance to be equal to those with a continuous correcting profile.

In addition to the work on reflecting surfaces, L. G. SCHULZ [29] has used the same techniques for correcting the spherical aberration of a lens. Both continuous profile and Fresnel correcting-plates were used and special techniques were evolved for obtaining thick deposits which did not show bloom, which latter sets a practical limit to the thickness of deposit which can be used. For lithium fluoride, the limit for a single layer is about 50,000 Å. The use of collodion separating layers enables this figure to be increased by a factor of three or four.

(b) *Apodization*

Diffraction sets an inescapable limit to the sharpness of the image formed by any optical system. The intensity distribution in, e.g., the image of the slit of a spectrometer consists of a central maximum bounded by alternate minima and maxima. The intensities of successive maxima fall away rapidly; those of the peaks adjacent to the central line are of little consequence except when spectral lines of widely different intensities occur close together. Under this condition, loss of contrast and general confusion may obtain.

The possibility of modifying the intensity distribution in the diffraction pattern from a grating by the use of irregular rulings was first discussed by R. W. DITCHBURN[30]. The term '*apodization*', referring to the process of suppressing the side maxima in an optical image was adopted by Jacquinot in 1946. This effect can be achieved by the introduction of a screen, in which either the absorption or the optical thickness varies in a calculable way over the surface, into the optical system to be apodized. The screens are conveniently made by thermal evaporation. The forms of screen required, with special reference to systems employing slits, are given in papers by B. DOSSIER, P. BOUGHON and P. JACQUINOT[31]. For a system using a slit source, the screen required must have a constant profile in the direction parallel to the slit. This requires that the plate on which the screen is to be deposited shall move at a constant speed, during the evaporation process, past the screen giving the required profile. P. JACQUINOT[32] uses for this purpose a freely-running carriage which is flipped to-and-fro electro-magnetically. The system is also useful for the preparation of linear wedges.

(c) *Thin films for anti-corrosion*

In many of the striking developments in multilayer techniques discussed in the early part of this chapter, little attention was paid to film properties other than those needed to fulfil the optical demands. A feature of considerable practical importance, if films are to be used outside the sacred precincts of the laboratory, is the resistance of such layers to corrosion by the malevolent elements of the ordinary atmosphere. The unsatisfactory nature of the compromise for producing a single-layer anti-reflecting film by evaporating CaF_2 in a poor vacuum is brought out when such a film is subjected to gentle abrasion. Although much less effective optically, magnesium fluoride is a more suitable material for blooming when stability and durability are considered.

In addition to their use in multilayer optical systems, evaporated films are of service purely in a protective capacity. Certain optical glasses have a poor resistance to atmospheric corrosion. The aluminium films used in reflecting telescopes are also found to deteriorate with time unless some protection is afforded. The rate of deterioration is particularly high in coastal neighbourhoods where a salt-laden atmosphere obtains.

The resistance to corrosion of films of several commonly-used dielectric film materials has been studied by F. FLAMANT[33]. As a quantitative measure of the deterioration of the film, the ratio of the light scattered to that reflected specularly was used. The test plate was placed in an integrating sphere, firstly so that it reflected an incident beam into a photocell and secondly, by tilting, so that the diffused light could be measured. The filmed surfaces were kept at 35° C over a trough of sea-water. This severe treatment served to destroy even the most effective of protecting films within a month or two.

Tests were made on films of aluminium fluoride, magnesium fluoride, silicon monoxide and silica. The fluorides were found to deteriorate extremely rapidly, leaving a surface which is inferior to that of the untreated surface after the same period. Magnesium fluoride held out better than the aluminium salt but was nevertheless completely destroyed in about 10 days. Silica was slightly better than the fluorides. Silicon monoxide was found to give the greatest degree of protection, breakdown being caused finally by the development of fissures through which attack of the underlying surface proceeded.

G. HASS and N. W. SCOTT[34] have studied the protection of aluminium mirrors both by silicon monoxide and by electrolytically-deposited aluminium oxide. The tests applied to the SiO-coated mirrors were (i) prolonged heating in air to 400° C and (ii) boiling in five per cent salt solution for one hour. SiO coating improves very considerably the performance of surfaces in these ordeals as well as increasing the resistance to abrasion of the surface. The absorption of the monoxide in the ultra-violet results in a poor reflectance for wavelengths below about 3500 Å. The effect is less severe for films deposited slowly and the absorption decreases on heating the film at 400° C in air, owing to the conversion of the monoxide into SiO_2. The very marked effect of the rate of deposition is shown in *Figure 7.26* where the properties of a film deposited at the rate of $\sim$100 Å/minute are compared with those for a rate of 800 Å/minute. Besides affording protection against atmospheric corrosion, silicon monoxide films are found to be of value if deposited on the mirror blank prior to the deposition of the aluminium film. An improved adhesion results.

Oxidizing an aluminium surface with the right type of anodizing agent—one which will produce a smooth, non-porous oxide film and which will not dissolve the film when formed—yields a surface which has a high resistance to abrasion and a fair resistance to contaminating atmospheres. Protection against the latter form of attack is inferior to that provided by silicon monoxide. Since the aluminium oxide film formed is highly transparent, even at wavelengths below 2500 Å, the optical performance of the aluminium is but slightly impaired. The anodizing arrangement used by Hass and Scott consists of a bath of 3 per cent tartaric acid to which ammonia is added to bring the pH to about 5·5. A pure aluminium plate serves as cathode and the surface to be treated is the anode. The thickness of the film formed depends on the voltage applied, the

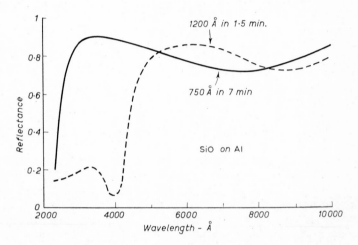

Figure 7.26 Influence of rate of deposition on the reflectance of SiO-coated aluminium

film growth being practically complete within 2–3 minutes. A thickness of 13 Å/volt is recorded for two minutes' anodizing, rising only to 13·5 Å/volt if the process is continued for fifteen minutes.

The oxide films so formed are structurally amorphous and free from pores. They may be easily removed by dissolving the underlying aluminium with dilute hydrochloric acid, when the film floats to the surface. They are mechanically very strong; films 100 Å thick may be picked up and cut out with scissors. They are very useful as specimen supports for electron diffraction or electron microscopy and may be heated to several hundred degrees centigrade without damage.

(d) *Photo-conductive cells, etc.*

On account of the difficulties presented by thin films in relation to theories of the solid state, little progress has been made in interpreting their electrical behaviour. Thin films of semi-conducting materials have an important practical application as photo-electric cells for the near infra-red region of the spectrum. Photo-conducting films of lead sulphide, lead selenide and lead telluride have enabled photo-electric measurements to be made at wavelengths out to nearly six microns. Used in conjunction with a chopped light source and tuned amplifier, the photo-conductive cell offers a more convenient and sensitive form of detection in this region than does the thermocouple.

Very little information is available on the design of photo-conductive cells, about which there appears to be an aura of alchemy. The thickness of the layer used is determined by the requirement that the fraction of the incident light which is absorbed shall be a maximum. From the equations developed in Section 4.4, it may be shown that, as the thickness of an absorbing film increases, the fraction of light absorbed increases to a maximum and then decreases as the bulk is approached.

An interesting and useful feature of films of stannous oxide is their good electrical conductivity combined with high optical transparency. They may be heated electrically and are used as coatings for glass surfaces which are required to be kept free from condensation. Gold films have been used for a similar purpose.

As mentioned at the beginning of this section, no more than a brief selection of the many and widespread applications of thin films has been given. The advances which have been made in recent years, both on the theoretical aspects of thin film optics and in the experimental techniques for producing the subtle combinations of layers which are now used have enabled great strides to be made in many fields. There is every reason to believe that this progress will continue apace.

·References

[1] ABELÈS, F. *J. de Phys.* **11** (1950) 310

[2] TURNER, A. F. *Ibid.* 444

[3] VAŠÍČEK, A. *Czech J. Phys.* **2** (1953) 71

[4] BANNING, M. *J.O.S.A.* **37** (1947) 792

[5] TURNER, A. F. *Bausch and Lomb Technical Report,* (Multilayer films) No. 5 (1951)

[6] RING, J. and WILCOCK, W. L. *Nature* **171** (1953) 648

[7] FRAU, D. C. *Rev. Opt.* **31** (1952) 161

[8] HADLEY, L. N. and DENNISON, D. M. *J.O.S.A.* **37** (1947) 451; *ibid.* **38** (1948) 483

[9] BILLINGS, B. H. *J. de Phys.* **11** (1950) 407

[10] KUHN, H. and WILSON, B. A. *Proc. Phys. Soc.* **B 63** (1950) 754

[11] BELK, J. A., TOLANSKY, S. and TURNBULL, D.   *J.O.S.A.* **44** (1954) 5
[12] TOLANSKY, S.   *Multiple-beam Interferometry*, O.U.P.
[13] DUFOUR, C.   *J. de Phys.* **11** (1950) 327
[14] — and HERPIN, A.   *Opt. Act.* **1** (1954) 1
[15] HEAVENS, O. S.   *J.O.S.A.* **44** (1954) 371
[16] POLSTER, H. D.   *Ibid.* **39** (1949) 1038
[17] GEE, A. E. and POLSTER, H. D.   *Ibid.* **39** (1949) 1044
[18] BILLINGS, B. H.   *Ibid.* **40** (1950) 471
[19] LYOT, B.   *C.R. Acad. Sci., Paris* **197** (1933) 1593
[20] BILLINGS, B. H., SAGE, S. and DRAISIN, W.   *Rev. Sci. Inst.* **22** (1951) 1009
[21] ABELÈS, F.   *J. de Phys.* **11** (1950) 403
[22] BANNING, M.   *J.O.S.A.* **37** (1947) 792
[23] DUFOUR, C.   *Le Vide* **No. 16–17** (1948) 480
[24] TURNBULL, D. T. and BELK, J. A.   *Lab. Practice* **1** (1952) 403
[25] GIACOMO, P. and JACQUINOT, P.   *J. de Phys.* **13** (1952) 59A
[26] GREENLAND, K. M. and BILLINGTON, C.   *Ibid.* **11** (1950) 418
[27] STRONG, J.   *Procedures in Experimental Physics*, Prentice-Hall, New York, 1945
[28] SCHULZ, L. G.   *J.O.S.A.* **37** (1947) 349
[29] —.   *Ibid.* **38** (1948) 432
[30] DITCHBURN, R. W.   *Proc. Roy. Irish Acad.* **44** (1938) 123
[31] DOSSIER, B., BOUGHON, P. and JACQUINOT, P.   *Journ. de Rech. du C.N.R.S.* **No. 11** (1950); *ibid.* **No. 12** (1950)
[32] JACQUINOT, P.   *J. de Phys.* **11** (1950) 361
[33] FLAMANT, F.   *Ibid.* 380
[34] HASS, G. and SCOTT, N. W.   *Ibid.* 394

AUTHOR INDEX

Abelès, F., 47, 61, 65, 70, 116, 119, 130, 209, 240
Alexandrow, A., 103
Andrade, E. N. da C., 25, 39
Antall, J. J., 152
Appleyard, E. T. S., 8, 36
Archard, J. F., 141
Ashworth, F., 43
Avery, D. G., 124, 132, 146–7, 172–3, 194–5, 197–8

Banning, M., 217, 241–3
Bär, W., 103
Belk, J. A., 226, 243
Betz, H., 103
Billings, B. H., 221–2, 235–6, 238
Billington, C., 248
Birks, L. S., 151–2
Blaisse, B. S., 69
Bor, J., 141
Bosworth, R. C. L., 102
Boughon, P., 252
Bradley, R. S., 99
Brattain, W. H., 137, 196–7
Briggs, H. B., 137, 196–7
Brockman, F. G., 101
Brossel, J., 107
Butler, C. C., 30, 164, 189

Chariton, J., 18
Clegg, P. L., 11, 98–99, 141, 165, 172–3, 184, 193–4, 198
Cochrane, W., 26
Cockcroft, J., 21–22, 25, 99
Collins, L. E., 9, 30, 38
Cotton, P., 89
Crittenden, E. C., 15
Crook, A. W., 59, 63, 98–99, 141

David, E., 139, 189–90
Dennison, D. M., 221
Ditchburn, R. W., 16, 18, 22, 39, 48, 252
Dixit, K. R., 26
Dossier, B., 252
Drude, P., 124–5, 140, 142–4, 150
Duffendack, O. S., 35–36
Dufour, C., 228, 230, 243

Eisenstein, A., 151
Essers-Rheindorf, G., 140, 145–6, 150
Estermann, K., 18, 25, 39, 99

Faust, R. C., 166–7
Finch, G. I., 37, 43
Fizeau, A., 106, 109–10, 113, 146, 225–6, 242
Flamant, F., 253
Fleischmann, R., 135–6
Fochs, P. D., 119–20, 130
Foley, F. M., 40
Försterling, K., 140, 144–5, 149, 194
Frau, D. C., 220
Frenkel, J., 18, 21, 38
Friedman, H., 151–2

Garnett, J. C. Maxwell, 169, 177, 182–3, 187–94, 198, 200–2
Gebbie, H. A., 137, 170, 194–8
Gee, A. E., 234, 236
Germer, L. H., 30
Giacomo, P., 245, 248
Goos, F., 181–3, 192–3
Greenland, K. M., 248
Güntherschulze, A., 103

Hadley, L. N., 221
Hallwachs, W., 162
Hambling, T. G., 38
Hammer, K., 158–9
Hanson, M., 125
Haringhuizen, P. J., 185
Harris, L., 14
Hass, G., 253–4
Hauschild, H., 143
Heavens, O. S., 9–10, 38, 110, 231
Herpin, A., 230
Hippel, A. von, 16
Holden, J., 107

Ingelstam, E., 110

Jacquinot, P., 245, 248, 252
Joffé, A., 103

Kent, C. V., 141
Kingdon, K. H., 100–1
Kinosita, H., 109
Knudsen, M., 18, 99
Kollmorgen, W., 208
König, H., 9
Krautkrämer, J., 163, 183–8, 191–2
Kuhn, H., 225

Langmuir, I., 100–1
Lawson, J., 141

Layton, D. N., 43
Leberknight, C. E., 140, 143, 150
Lennard-Jones, J. E., 26, 36
Levinstein, S., 9, 33–34, 37–38
Lovell, A. C. B., 8
Lustman, B., 140, 143, 150
Lyot, B., 236

McLauchlan, T. A., 113
Malé, D., 84, 86–87, 136–8, 149–50, 163, 165, 179, 181–2, 191–3
Malleman, R. de, 126
Martindale, J. G., 25
Mayer, H., 55, 103
Meier, W., 182–3, 195
Michelson, A., 119, 121, 130
Minor, R. S., 141
Müller, E. W., 42
Murmann, H., 136–7, 192

Newton, I., 113–14, 145–6

O'Bryan, H. M., 15, 194, 196–9
Olsen, L. O., 15
Omar, M., 151

Partzsch, A., 162
Pashley, D. W., 30
Picard, R. G., 35–36
Polster, H. D., 232, 234, 236

Rahbeck, H., 151
Ring, J., 220, 225, 229, 231
Rothen, A., 125, 141
Rouard, P., 63–64, 87, 118, 163–5, 169, 174–5, 182
Roulston, K. I., 22, 25
Rymer, T. B., 30, 38, 164, 189

Schopper, H., 92, 95, 104, 124, 133–6, 140, 149, 160, 177, 187–91, 193
Schulz, L. G., 30, 127, 130, 135, 146–7, 159, 199, 250–1

Schulze, R., 163–5, 182
Scott, G. D., 37–38, 98, 113, 163–6, 169, 172, 189, 201
Scott, N. W., 253–4
Semenoff, N., 18
Sennett, R. S., 37–38, 98, 113, 163–6, 166, 169, 172, 189, 201
Siegel, B. M., 14
Simon, I., 197
Smith, C. S., 15
Statescu, C., 101
Stewart, R. W., 121–2, 130
Strong, J., 14, 69, 250
Suhner, F., 126

Taylor, A. M., 141
Taylor, H. D., 208
Thomson, G. P., 26
Tolansky, S., 98, 105, 107, 113, 150, 226–7
Tousey, R., 132
Trompette, J., 163, 165
Tsien, H., 39
Turnbull, D., 226, 243
Turner, A. F., 211–12, 218, 221, 224, 231, 233, 237
Tyndall, J., 42

Vašíček, A., 63, 115, 124–6, 204, 214
Voigt, W., 141

Walkenhorst, W., 169
Weber, A. H., 152
Wiener, O., 105, 111
Wilcock, W. L., 220, 225, 229, 231
Wilson, B. A., 225
Winterbottom, A. B., 116, 140
Wolter, H., 139, 191, 193
Woltersdorff, W., 169–70
Wood, R. W., 18

Zehender, E., 40

SUBJECT INDEX

Absorbing media, 47 ff.
— propagation in, 47
— reflection at surface, 53–55
Absorption of films, 201–2
Adsorption, Frenkel theory, 18–22
— Lennard Jones, 36
Alkali halides, 33
— metals, 8
All-dielectric systems, 215
— in interferometry, 224–7
— interference filters, 228–31
Aluminium, 167–9, 198–200, 215
— fluoride, 209, 213, 253
— oxide, 213, 215, 253
Anti-corrosion films, 252
Antimony sulphide, 160–5, 217, 220
Anti-reflecting systems, 208 ff.
— angle of incidence, 214
— double film, 210–13
— single film, 209–10
— triple film, 213–14
Apodisation, 251–2
Applications of thin films, 207 ff.
Aspherical surfaces, 250–1

Barium, 198–9
Beam intensity, 9
Beryllium, 198–9

Cadmium, deposition, 39–42
Calcium, 198–9
— fluoride, 209, 220
— silicate, 213
Cerium, 198–9
Cold mirrors, 224
Complex index, 49
Contamination by source, 10
Copper, 195, 199–200
Critical beam intensity, 21
— temperature, 21
Cryolite, 209, 213, 220, 230

Deposition of films, 2, 6 ff.
— chemical, 17
— evaporation, 11 ff.
— electrolysis, 17
— sputtering, 16–17
Doubly refracting films, 92–95

Electrodeposited films, 43–44
Electron beam, 37–38
Evaporation plant, 13

Evaporation, sources, 11 ff.
— boat, 12
— conical basket, 15
— electron bombardment, 15
— helix, 14
— induction heating, 16
— oven, 11

Feco fringes, 105, 111–13
Field emission microscope, 42
Film thickness (non-optically), 96 ff.
— by capacitance, 103
— by electrolysis, 101
— by ionization, 100–1
— by molecular rays, 99
— by radioactive tracers, 152–3
— by resistance, 102
— significance, 96
— by weighing, 97–99
— X-ray methods, 151–2
Film thickness (optically), 103 ff.
— absorbing films, 131 ff.
— by Brewster angle, 119
— by colours, 113–16
— by light slit, 150–1
— Malé's method, 138–9
— by Michelson interferometer, 119–20
— by multiple-beams, 105 ff., 226
— by polarimetry, 123–7
— by reflectance, 116–18
— Schopper's method, 133–6
— summary, 130–1
Fizeau fringes, 105–11
— high-contrast, 225–7
Frenkel, theory of adsorption, 18–22
Fresnel coefficient, 52–53
— anisotropic film, 94
— by nomogram, 84
Frustrated total reflection filter, 231–5
— birefringent, 234–5

Germanium, deposition of, 38
— high-reflecting film, 217
— optical constants of, 195–8
— transmission of, 168
Gold films, colours of, 169, 176
— homogeneity of, 190–1
— optical constants of, 181 ff., 199-200
— reflectance of, 162–6
— transmittance of, 162–6
Graphical methods, 80 ff.
— general, 82–89
— transparent layers, 80–82

High-efficiency reflecting systems, 215 ff.
Homogeneity of films, 190–1, 203–5, 233–4

Interference filters, 227–31
— accuracy of layers, 230–1
— in infra-red, 229–30

Lanthanum, 198–9
Lead chloride, 213
Lithium fluoride, 209, 213

Magnesium, 198–9
— fluoride, 209–10, 213, 215, 253
Manganese, 198–9
Maxwell Garnett theory, 177–80
— Schopper's extension, 188–90
Maxwell's equations, 49–50
Metal+dielectric systems, 221–3
Monitoring, 242 ff.
— Giacomo & Jacquinot, 245–8
— Greenland & Billington, 248–9
— by intensity, 243–5
— visually, 243
Multiple layers, filters, 228–31, 235–8
— high-reflecting, 217–21
— reflectance of, 63 ff.
— transmittance of, 63 ff.
— use in interferometry, 224–7

Notation, 46–47

Optical constants, 49
— Avery, 146
— discussion of methods, 148–50
— Drude, 142–3
— Essers-Rheindorf, 145
— Försterling, 144–5
— Leberknight & Lustman, 143–4
— Malé, 138–9
— Murmann, 136–8
— results of measurements, 176 ff.
— Schopper, 133–6
— Schulz, 146–8
— very thin layers, 139, 142 ff.
Optical impedance, 66

Phase change, on reflection, 89–91
— on transmission, 90–92
— results, 170 ff.
Photoconductive cells, 254
Platinum, optical constants, 194–5
— phase change, 175
— reflectance, 161
— transmittance, 161
Polarimetric methods, experimental, 140–1
— for optical constants, 133 ff.
— for refractive index, 123 ff.
— for thickness, 123–7

Polarization, filter, 236–8
— by thin films, 238–41
Polarizing beam splitter, 241–2
Protective films, 252

q-value, 177–80, 183, 188, 200

Rate of evaporation, 9
Reflectance, absorbing medium, 54
— double-refracting film, 94
— from optical impedance, 66–69
— graphically, 80–89
— matrix method, 69 ff.
— multiple layers, 63 ff., 74–76
— results, 156 ff.
— single film, 58–59, 77–78
— total reflection, 232
— transparent media, 54
— two films, 66–69
Reflection coefficient (reflectance), 53
Refractive index, 113 ff.
— Abbe refractometer, 127–9
— Brewster angle, 119
— Drude, 124–5
— Malleman & Suhner, 126–7
— Michelson interferometer, 119–21
— non-uniform films, 121–3
— polarimetry, 123 ff.
— reflectance measurement, 116–18
— summary of methods, 130–1
— Vašíček, 125–6

Sensitization, 38 ff.
Silica, 253
Silicon monoxide, 253
Silver chloride, 213
Silver films, colours, 169, 176
— optical constants of, 192–4, 199–200
— phase change at, 172–3
— reflectance of, 166–7
— as reflectors, 215
Sodium, deposition of, 39
Sputtering, 16–17
Stacks, quarter-wave, 218–221
Strontium, 198–9
Structure of films, 24 ff.
— alkali halides, 33
— by electron diffraction, 26–30, 33–34
— by electron microscope, 31, 33–34
— by optical methods, 25–26
— of metals, 33–37
— silver, 166

Target surface, 9
Tellurium, 217, 230
Thermal evaporation, 7, 11 ff.
Tin, deposition of, 40
— optical constants of, 198
Titanium dioxide, 217

260

Transmittance, 49 ff.
— doubly-refracting film, 94
— graphically, 87–88
— matrix method, 69 ff.
— multiple films, 63 ff.
— results, 156 ff.

Transmittance, single film, 55–61, 77–78
— at total reflection, 232
— two films, 62–63, 79–80

Zinc, deposition of, 40–42
— sulphide, 158, 215, 217, 220

DATE DUE